T0254986

Lecture Notes in Computer Science 13581

Founding Editors

Gerhard Goos
Karlsruhe Institute of Technology, Karlsruhe, Germany

Juris Hartmanis
Cornell University, Ithaca, NY, USA

Editorial Board Members

Elisa Bertino
Purdue University, West Lafayette, IN, USA

Wen Gao
Peking University, Beijing, China

Bernhard Steffen
TU Dortmund University, Dortmund, Germany

Moti Yung
Columbia University, New York, NY, USA

More information about this series at https://link.springer.com/bookseries/558

Sharib Ali · Fons van der Sommen ·
Bartłomiej Władysław Papież ·
Maureen van Eijnatten · Yueming Jin ·
Iris Kolenbrander (Eds.)

Cancer Prevention Through Early Detection

First International Workshop, CaPTion 2022
Held in Conjunction with MICCAI 2022
Singapore, September 22, 2022
Proceedings

Springer

Editors
Sharib Ali 🄳
University of Leeds
Leeds, UK

Fons van der Sommen 🄳
Eindhoven University of Technology
Eindhoven, The Netherlands

Bartłomiej Władysław Papież 🄳
University of Oxford
Oxford, UK

Maureen van Eijnatten 🄳
Eindhoven University of Technology
Eindhoven, The Netherlands

Yueming Jin 🄳
University College London
London, UK

Iris Kolenbrander 🄳
Eindhoven University of Technology
Eindhoven, The Netherlands

ISSN 0302-9743   ISSN 1611-3349 (electronic)
Lecture Notes in Computer Science
ISBN 978-3-031-17978-5   ISBN 978-3-031-17979-2 (eBook)
https://doi.org/10.1007/978-3-031-17979-2

© The Editor(s) (if applicable) and The Author(s), under exclusive license
to Springer Nature Switzerland AG 2022
This work is subject to copyright. All rights are reserved by the Publisher, whether the whole or part of the material is concerned, specifically the rights of translation, reprinting, reuse of illustrations, recitation, broadcasting, reproduction on microfilms or in any other physical way, and transmission or information storage and retrieval, electronic adaptation, computer software, or by similar or dissimilar methodology now known or hereafter developed.
The use of general descriptive names, registered names, trademarks, service marks, etc. in this publication does not imply, even in the absence of a specific statement, that such names are exempt from the relevant protective laws and regulations and therefore free for general use.
The publisher, the authors, and the editors are safe to assume that the advice and information in this book are believed to be true and accurate at the date of publication. Neither the publisher nor the authors or the editors give a warranty, expressed or implied, with respect to the material contained herein or for any errors or omissions that may have been made. The publisher remains neutral with regard to jurisdictional claims in published maps and institutional affiliations.

This Springer imprint is published by the registered company Springer Nature Switzerland AG
The registered company address is: Gewerbestrasse 11, 6330 Cham, Switzerland

Preface

CaPTion 2022, the 1st International Workshop on Cancer Prevention through early detecTion, was organized as a satellite event of the 25th International Conference on Medical Image Computing and Computer Assisted Intervention (MICCAI 2022) in Singapore. The main idea of founding CaPTion was to create a new research interface where medical image analysis, machine learning, and clinical researchers could interact and address the challenges related to early cancer detection using computational methods.

Early cancer diagnosis and its treatment for the long-term survival of cancer patients have been a battle for decades. In 2020, 19.3 million cancer cases and almost 10 million deaths were reported, with lung (18%), colorectal (9.4%), liver (8.3%), stomach (7.7%), and female breast cancer (6.9%) being the leading causes of mortality. While computational methods in medical imaging have enabled the detection and assessment of cancerous tumors and assist in their treatment, early detection of cancer precursors opens an opportunity for early treatment and prevention. The workshop provided an opportunity to present research work in medical imaging around the central theme of early cancer detection. It strived to address the challenges that must be overcome to translate computational methods to clinical practice through well-designed, generalizable, interpretable, and clinically transferable methods. Through this new workshop, we aimed to identify a new ecosystem that would enable comprehensive method validation and reliability of methods, setting up a new gold standard for sample size and elaborating evaluation strategies to identify failure modes of methods when applied to real-world clinical environments.

The CaPTion 2022 proceedings contain 16 high-quality papers of 8 to 11 pages preselected through a rigorous peer review process (with an average of three reviews per paper). All submissions were peer-reviewed through a double-blind process by at least two members of the scientific review committee, comprising 21 experts in the field of medical imaging, especially within early cancer detection. The accepted manuscripts cover various medical image analysis methods and applications. In addition to the papers presented in this LNCS volume, the workshop hosted three keynote presentations from world-renowned experts: Fergus Gleeson (University of Oxford), Kristy K. Brock (University of Texas MD Anderson Cancer Center), and Michael Byrne (Vancouver General Hospital/University of British Columbia).

We wish to thank all the CaPTion 2022 authors for their participation and the members of the scientific review committee for their feedback and commitment to the workshop. We are very grateful to our sponsors: the NIHR Biomedical Research Centre, UK, and Satisfai Health Inc, Canada, for their valuable support.

The proceedings of the CaPTion workshop are published as a separate LNCS volume in conjunction with MICCAI 2022.

September 2022   Sharib Ali
                                              Fons van der Sommen
                                             Maureen van Eijnatten
                                             Bartłomiej W. Papież
                                                    Yueming Jin
                                                 Iris Kolenbrander

Organization

Program Committee Chairs

Sharib Ali — University of Leeds, UK
Fons van der Sommen — Eindhoven University of Technology, The Netherlands
Maureen van Eijnatten — Eindhoven University of Technology, The Netherlands
Bartłomiej W. Papież — University of Oxford, UK
Yueming Jin — University College London, UK
Iris Kolenbrander — Eindhoven University of Technology, The Netherlands

Scientific Review Committee

Christian Daul — Université de Lorraine, France
Mariia Dmitrieva — University of Oxford, UK
Haoyu Dong — Duke University, USA
Xiaohong Gao — Middlesex University, UK
Noha Ghatwary — University of Lincoln, UK
Soumya Gupta — University of Oxford, UK
Yang Hu — University of Oxford, UK
Benoit Huet — Median Technologies, France
Dimitris Iakovidis — University of Thessaly, Greece
Debesh Jha — Northwestern University, USA
Fucang Jia — Chinese Academy of Sciences, China
Carolus Kusters — Eindhoven University of Technology, The Netherlands
Luyang Luo — The Chinese University of Hong Kong, Hong Kong
Gilberto Ochoa-Ruiz — Tecnologico de Monterrey, Mexico
Nishant Ravikumar — University of Leeds, UK
Duygu Sarikaya — Gazi Üniversitesi, Turkey
Adam Szmul — University College London, UK
Mansoor Ali Teevno — Tecnologico de Monterrey, Mexico
Ziang Xu — University of Oxford, UK
Le Zhang — University of Oxford, UK
Yifan Zhang — Vanderbilt University, USA

Contents

Classification

3D-Morphomics, Morphological Features on CT Scans for Lung Nodule Malignancy Diagnosis

Elias Munoz, Pierre Baudot[✉], Van-Khoa Le, Charles Voyton,
Benjamin Renoust, Danny Francis, Vladimir Groza, Jean-Christophe Brisset,
Ezequiel Geremia, Antoine Iannessi, Yan Liu, and Benoit Huet

Median Technologies, 1800 Rte des Crêtes Batiment B, 06560 Valbonne, France
pierre.baudot@mediantechnologies.com

Abstract. Pathologies systematically induce morphological changes, thus providing a major but yet insufficiently quantified source of observables for diagnosis. The study develops a predictive model of the pathological states based on morphological features (3D-morphomics) on Computed Tomography (CT) volumes. A complete workflow for mesh extraction and simplification of an organ's surface is developed, and coupled with an automatic extraction of morphological features given by the distribution of mean curvature and mesh energy. An XGBoost supervised classifier is then trained and tested on the 3D-morphomics to predict the pathological states. This framework is applied to the prediction of the malignancy of lung's nodules. On a subset of NLST database with malignancy confirmed biopsy, using 3D-morphomics only, the classification model of lung nodules into malignant vs. benign achieves 0.964 of AUC. Three other sets of classical features are trained and tested, (1) clinical relevant features gives an AUC of 0.58, (2) 111 radiomics gives an AUC of 0.976, (3) radiologist ground truth (GT) containing the nodule size, attenuation and spiculation qualitative annotations gives an AUC of 0.979. We also test the Brock model and obtain an AUC of 0.826. Combining 3D-morphomics and radiomics features achieves state-of-the-art results with an AUC of 0.978 where the 3D-morphomics have some of the highest predictive powers. As a validation on a public independent cohort, models are applied to the LIDC dataset, the 3D-morphomics achieves an AUC of 0.906 and the 3D-morphomics+radiomics achieves an AUC of 0.958, which ranks second in the challenge among deep models. It establishes the curvature distributions as efficient features for predicting lung nodule malignancy and a new method that can be applied directly to arbitrary computer aided diagnosis task.

Keywords: Mesh · Radiomics · Computed tomography · Lung cancer screening · Computer aided diagnosis · Computational anatomy

Supplementary Information The online version contains supplementary material available at https://doi.org/10.1007/978-3-031-17979-2_1.

© The Author(s), under exclusive license to Springer Nature Switzerland AG 2022
S. Ali et al. (Eds.): CaPTion 2022, LNCS 13581, pp. 3–13, 2022.
https://doi.org/10.1007/978-3-031-17979-2_1

1   Introduction

Since the beginning of medicine, morphological characteristics have provided phenotypes of symptoms allowing to establish the clinical diagnosis. According to the generic principle of a correspondence between structure and function in biology, any dysfunction of a biological process goes in hand with a pathological deformation of the underlying biological structure. Unfortunately, morphological features are often difficult to quantify; at best limited to coarse quantifiers, if not purely qualitative observations. As a consequence, most of the observations of morphological features are not reliable in clinical practice, and are progressively replaced by biomarkers. The aim of this study is twofold: first, the development of a basic and generic method of morphological classification based on the extraction of mesh curvature distribution; second, a validation of the method on clinical prediction of lung nodules malignancy. Curvature is one of the main topological invariant, in the sense of the Bonnet-Gauss theorem, and hence a robust descriptor of the shape of manifold. However, in order to obtain feature that are sensible to deformations and hence not diffeomorphic invariant, we consider the distribution of curvatures which encodes a wide spectrum of deformations. For a theoretical context, we refer to Federer [14] and note that we consider the Bonnet-Gauss theorem as giving a signed measure, which distribution of its atomic element (at each vertex) defines our set of features. Non-invasive automatic diagnosis of disease on CT images are based on different kinds of models illustrated here in the context of lung cancer: (1) size features-based models (basic morphology) as in the popular LungRADS model [22], (2) clinical features-based model like PANCAN [26] that mostly relies on patients' clinical and historical information like age, antecedent and smoking habits sometime associated with size [16], (3) radiomics features-based models, commonly textural and image statistics features [3,25,28], and (4) deep network models such as convolutions nets [7] or attention-based models [6], that provide the best performances and are theoretically optimal. Radiomics most often rely on textural features, luminance intensity statistics features, and in some case to basic shape quantification such as 2D ellipse's radius. In the context of lung cancer, the clinician proposes his diagnosis of nodule mostly based on the size of the nodule, along with two sets of qualitative features representing the shape of the nodule (margins, contours) and the luminance profile of the nodule (calcification, attenuation) [13]. Notably, spiculations and lobulations, which are spikes or bumps on the surface of the pulmonary nodules, as opposed to the "sphericity", are important predictors of the lung nodule malignancy. In an unsupervised sub-classification study of lung or head-and-neck tumors, Aerts et al. [3] designed a state-of the-art library of radiomics including image intensity, texture, but also introduced shape features like compactness or sphericity computed as functions of surface to volume ratios [15], along with the more classical size and volume features. Their sets of 110 features will be used as a baseline of the present study. More refined morphological study is realized by Leonardi et al. [20], that proposes to localize kidney exophytic tumors on a surface mesh using a maximum curvature or a recursive labelling of the vertices method. Also remarkably, Choi et al. quantify spiculation by reconstructing a 3D mesh from a mask of a nodule, and conformally map it to a sphere and then measure the area distortion [12]. Spiculations may be identified by the set of minima of area distortion. Completing this spiculation feature with size, attachment, and lobulation features, the authors obtained an AUC of 0.82 with a training and test on LIDC dataset.

2 Methods

2.1 Data Sets

NLST Dataset. The National Lung Screening Trial (NLST) is a US cancer screening program with 7 years follow-up study with yearly survey [1,17], with at most the first 3 years CT, available publicly on request. We only consider the 618 cancer patients, and 1201 non-cancer patients with nodules taken randomly among the 8210 with 3 time points eligible for download, with Low Dose CT scans. For each time points, a single CT scan is selected among the multiples CT scans kernels available using the same criterion as Ardila et al. [7]. The annotations, conducted by two expert radiologists, consisted in two tasks: (1) a semi-manual segmentation of all segmentable nodules and (2) a disambiguation task that associate each segmented nodule with its NLST GT that contains notably the nodules result of biopsy and localization, and to each new detected lesion not in NLST GT with a new GT. 523 cancer patients have at least one identified malignant lesion identified by biopsy. Only solid and part-solid parenchymal nodules are selected. The set of malignant lesions is defined as the lesion identified by biopsy as malignant at the time of the cancer diagnosis. The set of benign lesions is defined as all the selected lesion at all time points of non-cancer patients, and the set of calcified lesions in cancer patients that are known to be benign nodules. The remarkable aspect of this database relies in the quality of the malignancy GT: the malignancy of lesions is confirmed at the histological level by invasive procedure. The non-cancer status of patients and nodules is confirmed by up to 7 years follow up. This selection hence includes all the False Positive (FP) of clinician diagnosis (biopsied lesions confirmed benign by histology) as well as all False Negative (FN) of clinicians (not biopsied lesions). The data set is split randomly at the patient level into a train+validation set and a test set containing 168 cancer and 330 benign patients. The resulting number of nodules of each set is resumed in Table 1: the lesion inclusion-exclusion resulted into a total of 523 malignant lesions and 15726 benign lesions.

LIDC-IDRI Independent Test Cohort. The Lung Image Database Consortium (LIDC-IDRI) database is a public and multicentric US lung cancer screening database [8] of 1018 CT scans from 1010 patients. For each patient, the dataset includes a CT scan, the annotations and segmentations performed by up to height radiologists. As achieved by previous studies [6,27,30] we only consider the nodules annotated by at least 4 radiologists with a diameter greater or equal than 3 mm. The malignancy of each nodule was rated from 1 to 5 by the radiologist. Following [5,6], the final malignancy labels is obtained by taking the median value of the ratings of all radiologists and the nodules with a median of 3 were excluded as no benign or malignant assignment can be given for them. This selection results in a total of 656 nodules, among which 352 are malignant (54%), and 304 are benign (46%) (cf. Table 1). As in [18,27], a 50 % consensus criterion is opted to generate the segmentation mask boundaries of the nodules in order to remove the variability among the radiologists.

Table 1. Numbers of lesions for each set and subsets of the study tables.

Dataset	# benign nodule	# malignant nodule	total # nodule
NLST train-validation set	11304	372	11676
NLST test set	4422	151	4573
NLST total	15726	523	16249
LIDC independent test	304	352	656

2.2   Data Analysis Models

The 3D-morphomics process reconstructs a 3D mesh from a binary mask of an organ or sub-tissue, and automatically extract curvature features from the mesh. The features are then used for the diagnosis, e.g. the prediction of the malignancy of lung nodules. Thus, the process is composed of 3 main steps (Fig. 1).

Pre-processing: The slice spacing vary according to center and CT acquisition apparatus, for example in the NLST and LIDC collections, ranging from 0.45 to 5 mm. Masks and CT volumes are thus re-sampled to cubical voxel size 0.625^3 mm using nearest-neighbor method and cubical spline respectively. Patches of $64 \times 64 \times 64$ voxels centered on the barycenter of mask are then extracted.

3D-Morphomics Features: A mesh is constructed from the boundaries of a 3D arbitrary mask volume of an organ using the Marching Cube of Lewiner algorithm as it resolves ambiguities and guarantees topologically correct results. The reconstructed mesh is then simplified to improve triangles quality by collapsing short edges, splitting long edges, removing duplicated vertices, duplicated faces,

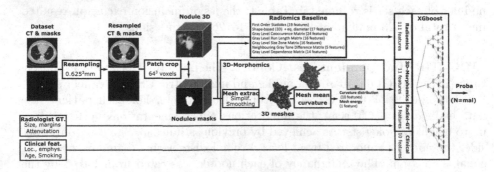

Fig. 1. The workflow of 3D-morphomics

and zero area faces. As the voxels resolution induce noise in the estimation of the curvature, we tested Poisson's, Taubin's, and z axis anisotropic smoothing of the meshes. None of them resulted in performance improvements and they were hence discarded. Both Gauss or mean discrete curvature are extracted [23]. The mean curvature K_i at a vertex i is defined via the Steiner approximation [23] $K_i = \frac{1}{4} \sum_{ij} \theta_{ij} l_{ij}$, where θ_{ij} is the edge dihedral angle and l_{ij} is the edge length as defined above (and the sum is taken over halfedges extending from i). The probability distribution (normalized to unit area) of the mean curvature K_G on the mesh are computed and binned to 10 range values distribution histograms (with fixed max and min of -0.2 and 0.2), together with the associated mesh's energy $E(M) = \int_M |K_G| dM$. The windowing and number of bins of the distribution constitute the main hyper-parameters of the method that can be tuned as classically. It provides the 11 morphological features that are used for the classification. The number of bins shall be chosen as a function of the sample size in order to avoid curse of dimensionality problems.

Radiomics: Following [3,15], 111 radiomic features are extracted, including, 19 of first order statistics, 17 3D shape-based (one additional redundant feature of "volume" given computed by counting the voxels volumes), 24 Gray Level Cooccurence Matrix (GLCM), 16 Gray Level Run Length Matrix (GLRLM), 16 Gray Level Size Zone Matrix (GLSZM), 5 Neighbouring Gray Tone Difference Matrix (NGTDM), 14 Gray Level Dependence Matrix (GLDM).

Clinical Features: they consist in patient information provided by NLST database: age ('age'), pack year ('pkyr'), Average number of cigarettes per day ('smokeday'), Average number of cigarettes per day, Cigarette smoking status ('cigsmok', current vs former), Age at smoking onset ('smokeage'), Total years of smoking ('smokeyr'), family antecedent ('family_ant', if lung cancer was diagnosed in brother (+1), sister (+1), mother (+1), father (+1)), localization of the nodule on the height axis ('localization', lobe: down, middle, up), Emphysema PSE ('emph_PSE', no, mild, substantial), Emphysema score ('emph_sc', 0–4).

Radiologist GT Features: They consist in the longest axis diameter measured by our radiologist using semi-automated tool, and two qualitative features provided by NLST radiologists: the "Margins" (smooth, poorly defined, spiculated), and "Attenuation" (mixed, Ground Glass, soft tissue, fat, water).

Classifier: The inference model is a gradient boosted decision tree (XGBoost [10]). The hyper-parameters of XGBoost are tuned on the train-validation set of NLST with a 80-20% split using Bayesian descent on a binary logistic loss. The space of exploration with their range is: 'max_depth' [3–18], 'gamma' [0,9], 'reg_alpha' [1e−5, 1e−4, 5e−4, 1e−3, 5e−3, 1e−2, 5e−2, 0.1, 0.5, 0.7, 1, 2, 3, 5, 10, 20, 50, 80, 100], 'reg_lambda' [0,1], 'colsample_bytree' [0.5,1], 'min_child_weight' [0–10], 'subsample' [0.5,1] with a number of estimator set to 180. The model is retrained on the train set with the set of optimal parameters with a learning rate of 0.01 and 1000 estimators. The input imbalance (scale_pos_weight) was set to the observed imbalance. The models are then validated against the independent test set of NLST or the LIDC cohort, and models

AUC are compared using unpaired 2-sided Welch t-test $p < 0.05$ on 5000 bootstrap samples of the test sets.

Dependencies and Computation Time: PyRadiomics 3.0.1, Scikit-Learn 0.24.2, Pymesh 0.3, Xgboost 1.5.2, OpenMesh, Scikit-Image 0.18.3, hyperopt 0.2.7, SimpleITK 2.2.1. The computation of 3D-morphomics is light and takes 1.2 s per nodule on average on a standard CPU (IntelCore i7-6700K CPU@4.00 GHz).

3   Results

3.1   3D-Morphomics

A visualisation of examples and basic statistical analysis of the curvature distribution features is illustrated in Fig. 2 for the whole set of nodules of NLST dataset. As expected, the curvature distributions obtained from malignant and benign nodules are very different, the malignant nodules displaying a clear over-representation of high negative and positive curvature values, while benign nodules display a clear over-representation of low curvature values. This shows that the computed features effectively captures the shape irregularities such as the spiculations that are discriminative features of malignancy for clinician. Considering each curvature features independently, the mean of the distributions of the malignant and benign are highly significantly different for each feature (for all features we have $p << 0.001$ with a two-sided unequal variance t-test on the mean value). It can also be observed from Fig. 2c. That in both cases the distributions have an overall bias toward positive curvatures, as a direct consequence of the fact that nodules are isomorphic to the sphere. The study of Gaussian curvatures is not shown here and, as illustrated by Leonardi [19] (p.100) combining Gaussian and mean curvatures allows to encode more information about the shape such as valleys, ridges, saddles, pick and peat, hence suggesting some possible improvements on the present features for more subtle classifications.

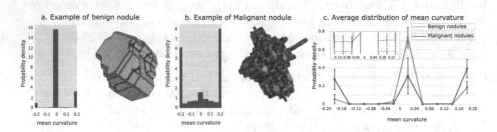

Fig. 2. Examples of Benign and Malignant nodules curvature distribution and meshes (a., b., the mean curvature values at each vertex are color coded from blue-negative, green-close to zero, red-positive) and (c.) the average distributions of mean curvature across all NLST set of malignant and benign nodules with error bars representing their standard deviations (enlargement for small values is given inset). (Color figure online)

Table 2. Models performances on NLST and LIDC test sets: AUC, sensitivity and specificity at the maximum of Youden index, accuracy, 5000 bootstraps mean AUC ± standard deviation and 95% Confidence Interval (bold: best automated models).

Models (test set)	AUC	Sens.	Spec.	Acc.	AUC Mean ± Stdev [95%CI]
3D-morphomics (NLST)	0.964	90.7	91.1	0.94	0.964 ± 0.006 [0.952 0.976]
Radiomics (NLST)	0.976	92.7	93.6	0.94	0.977 ± 0.004 [0.968 0.984]
Clinical feat. (NLST)	0.58	64.2	52.0	0.61	0.58 ± 0.023 [0.534 0.625]
Radiologist GT feat. (NLST)	0.979	93.3	93.4	0.93	0.979 ± 0.004 [0.971 0.986]
Brock NLST model [29] (NLST)	0.826	72.0	82.0	0.81	0.826 ± 0.02 [0.786 0.864]
3D-morphomics+Radiomics (NLST)	**0.978**	92.7	94.7	**0.95**	**0.979 ± 0.004 [0.971 0.986]**
3D-morphomics (LIDC)	0.906	84.0	85.8	0.84	0.906 ± 0.012 [0.883 0.928]
Radiomics (LIDC)	0.956	88.0	90.7	0.89	0.956 ± 0.007 [0.941 0.969]
3D-morphomics+radiomics (LIDC)	0.958	91.7	87.1	0.90	0.958 ± 0.007 [0.944 0.971]

3.2 Lung Nodule Diagnosis Performances of 3D-Morphomics

Following the process exposed in Sect. 3.1, we trained 5 different models on the NLST train-valid set corresponding to the set of 3D-morphomics, radiomics, clinical, radiologist GT, and the combination of 3D-morphomics with radiomics features. The performances estimated on the NLST test set and on the LIDC dataset are resumed in Table 2. The 3D-morphomic model exhibits satisfying performances with an AUC of 0.964, in the sens that it significantly outperform the NLST Brock model [29] (Welch t-test n = 5000, $p < 0.05$), and indirectly the AUC of 0.82 obtained by Choi et al. using morphological features on LIDC [12]. The high positive curvature have the highest predictive power. The importance of features for each model is given in supplementary material. As expected, the radiomics model, which gathers 111 features representing size, classical 3D shape and luminance patterns functions (first order stat, GLCM, GLRLM, GLSZM, NGTDM, GLDM) provides even higher performances with an AUC of 0.976. Four among the 6 highest predictive features are 2D shape features, while the 2 left are NGTDM and GLDM features. The clinical model with features partially overlapping with the PANCAN model here applied at the nodule level, provides poor performance, indicating that clinical informations such as age, smoking habits, family antecedent, emphysema, have low impact on the prediction of a lesion malignancy. The nodule horizontal axis localization has far the most important predictive power, followed by family antecedent, the number of smoking years, and then age. As expected also, the radiologist GT model, based on 3 main diagnostic criterions of size, qualitative margins (shape) and attenuation (texture) of a nodule, gives a high AUC of 0.979. As a surprise, attenuation and margins had about twice greater predictive power than the size (given by the longest diameter). The main result of the paper is that the combination of radiomics and 3D-morphomics gives a performance of 0.978, as illustrated in Fig. 3. As shown in the graphic of feature importance, 6 curvature distribution features resides among the 30 features with highest predictive

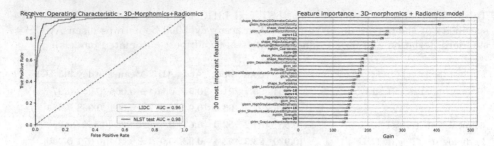

Fig. 3. Main results: (left) ROC Curves of the 3D-Morphomics+Radiomics model on NLST test set and LIDC dataset. (right) XGBoost feature importance (30 highest) on NLST test set. Curvature distribution features are in bold.

power. Within the top 5 of features with highest predictive power we can observe 2 shape feature (2D size and volume) 2 texture feature (grey level non uniformity of GLSZM and GLDM) and a 3D-morphomics feature of high positive curvature. We also tested the Brock NLST model [29], which combines some clinical and annotators GT features into a logistic model, and obtained an AUC of 0.826, lower than the originally reported but partly confirming recent results [11]. In order to validate the model and results on a publicly available independent cohort, the 3D-morphomics and 3D-morphomics+radiomics model are tested on the LIDC dataset, giving an AUC of 0.906 which is higher than the 0.82 obtained with other morphological feature by Choi et al. [12] and of 0.958 respectively (cf. Table 2). The latter score is ranking second at the associated LIDC characterization challenge, just after [6], and despite the clear disadvantage that our model was not trained on LIDC as opposed to the other solutions. The result of unpaired 2-sided Welch t-test ($n = 5000$, $p < 0.05$) rejects the null hypothesis that 3D-morpho+Radiomics and Radiomics have equal mean on both NLST and LIDC test set. We conclude that the good performance of the 3D-morphomics+radiomics model generalize satisfyingly to other data, and is robust to important dataset variations: while NLST is a low dose CT dataset, LIDC is not. We remark that the malignancy GT of LIDC is much more uncertain with respect to the true malignancy status of a nodule than the accurate estimation given by the biopsy results for malignancy and by the up to 7 year follow up for benignity. Notably, the malignancy GT of LIDC contains all the FP and FN of radiologists, e.g. the False Discovery Rate of radiologists on our NLST dataset is 5.9%. Hence, the results obtained on the test set of NLST and trained on NLST against the biopsy derived GT, gives a more accurate estimation on the real performances of our model than the results obtained with the LIDC set that are biased by radiologist diagnostic errors.

4   Conclusions

It is a surprise that morphomics and radiomics approaches, relying on predesigned features, can provide equivalent or even out-perform deep network models on a medical diagnosis task [2,4,21]. Given previously reported high performances of deep models on the NLST cohort [9], those results can only leave the hope that a combination of deep and 3D-morphomics solutions will achieve even better performances. This work establishes the distribution curvature as efficient features for predicting lung nodule malignancy and that the nodules shape have a predominant predictive power beyond the classical size used in everyday clinical practice at the image of the Lung-RADS score. Moreover, the method presented here based on a new family of observable can be applied directly to arbitrary computer aided diagnosis task, and is currently applied to liver fibrosis stage diagnostic where the nodularity of the liver surface is symptomatic. The model is quite generic in order to detect morphological deformations due to pathologies, but suffers from reducing the morphology to a 1D vector of curvature distribution, which erase some 2D spatial information. To overcome this limitation, an appealing perspective could be to use Graphical Neural Network classifiers taking as input the 2D network of the meshes, as explored in [24].

References

1. Reduced lung-cancer mortality with low-dose computed tomographic screening. N. Engl. J. Med. **365**(5), 395–409 (2011). https://doi.org/10.1056/NEJMoa1102873
2. Lung nodule classification on LIDC-IDRI challenge (2022). https://paperswithcode.com/sota/lung-nodule-classification-on-lidc-idri
3. Aerts, H.J., et al.: Corrigendum: decoding tumour phenotype by noninvasive imaging using a quantitative radiomics approach. Nat. Commun. **5**(1), 4006 (2014). https://doi.org/10.1038/ncomms5006
4. Al-Shabi, M., Lan, B.L., Chan, W.Y., Ng, K.H., Tan, M.: Lung nodule classification using deep local-global networks. Int. J. Comput. Assist. Radiol. Surg. **14**(10), 1815–1819 (2019)
5. Al-Shabi, M., Lee, H.K., Tan, M.: Gated-dilated networks for lung nodule classification in CT scans. IEEE Access **7**, 178827–178838 (2019). https://doi.org/10.1109/ACCESS.2019.2958663
6. Al-Shabi, M., Shak, K., Tan, M.: ProCAN: progressive growing channel attentive non-local network for lung nodule classification. Pattern Recogn. **122**, 108309 (2022). https://doi.org/10.1016/j.patcog.2021.108309
7. Ardila, D., et al.: End-to-end lung cancer screening with three-dimensional deep learning on low-dose chest computed tomography. Nat. Med. **25**(6), 954–961 (2019). https://doi.org/10.1038/s41591-019-0447-
8. Armato, S.G., et al.: The lung image database consortium (LIDC) and image database resource initiative (IDRI): a completed reference database of lung nodules on CT scans. Med. Phys. **38**(2), 915–931 (2011). https://doi.org/10.1118/1.3528204
9. Baudot, P., et al.: Development and validation of a machine learning based CADx designed to improve patient management in lung cancer screening programs. In: Proceedings of ECR 2022, Vienna, July 2022
10. Chen, T., Guestrin, C.: XGBoost: a scalable tree boosting system. In: Proceedings of the 22nd ACM SIGKDD International Conference on Knowledge Discovery and Data Mining, New York, NY, USA, pp. 785–794 (2016). https://doi.org/10.1145/2939672.2939785

11. Chetan, M., Dowson, N., Price, N., Ather, S., Nicolson, A., Gleeson, F.: Developing an understanding of artificial intelligence lung nodule risk prediction using insights from the Brock model. Eur. Radiol. (2022). https://doi.org/10.1007/s00330-022-08635-4
12. Choi, W., Nadeem, S., Alam, S.R., Deasy, J.O., Tannenbaum, A., Lu, W.: Reproducible and interpretable spiculation quantification for lung cancer screening. Comput. Methods Programs Biomed. **200**, 105839 (2021). https://doi.org/10.1016/j.cmpb.2020.105839
13. Erasmus, J.J., Connolly, J.E., McAdams, H.P., Roggli, V.L.: Solitary pulmonary nodules: part I. Morphologic evaluation for differentiation of benign and malignant lesions. RadioGraphics **20**(1), 43–58 (2000). https://doi.org/10.1148/radiographics.20.1.g00ja0343
14. Federer, H.: Curvature measures. Trans. Am. Math. Soc. **93**(3), 418–491 (1959). https://doi.org/10.2307/1993504
15. van Griethuysen, J.J., et al.: Computational radiomics system to decode the radiographic phenotype. Can. Res. **77**(21), 104–107 (2017). https://doi.org/10.1158/0008-5472.CAN-17-0339
16. Huang, P., et al.: Prediction of lung cancer risk at follow-up screening with low-dose CT: a training and validation study of a deep learning method. Lancet Digital Health **1**(7), e353–e362 (2019). https://doi.org/10.1016/S2589-7500(19)30159-1
17. Jemal, A., Fedewa, S.A.: Lung cancer screening with low-dose computed tomography in the United States-2010 to 2015. JAMA Oncol. **3**(9), 1278–1281 (2017). https://doi.org/10.1001/jamaoncol.2016.6416
18. Kubota, T., Jerebko, A.K., Dewan, M., Salganicoff, M., Krishnan, A.: Segmentation of pulmonary nodules of various densities with morphological approaches and convexity models. Med. Image Anal. **15**(1), 133–154 (2011). https://doi.org/10.1016/j.media.2010.08.005
19. Leonardi, V.: Modélisation dynamique et suivi de tumeur dans le volume rénal. These de doctorat, Aix-Marseille, November 2014. http://www.theses.fr/2014AIXM4056
20. Leonardi, V., Vidal, V., Daniel, M., Mari, J.-L.: Multiple reconstruction and dynamic modeling of 3D digital objects using a morphing approach. Vis. Comput. **31**(5), 557–574 (2014). https://doi.org/10.1007/s00371-014-0978-6
21. Al-Shabi, M., Lee, H.K., Tan, M.: Gated-dilated networks for lung nodule classification in CT scans. IEEE Access **7**, 178827–178838 (2019)
22. McKee, B.J., Regis, S.M., McKee, A.B., Flacke, S., Wald, C.: Performance of ACR lung-RADS in a clinical CT lung screening program. J. Am. Coll. Radiol. **12**(3), 273–276 (2015). https://doi.org/10.1016/j.jacr.2014.08.004
23. Meyer, M., Desbrun, M., Schröder, P., Barr, A.H.: Discrete differential-geometry operators for triangulated 2-manifolds. In: Hege, H.C., Polthier, K. (eds.) Visualization and Mathematics III, pp. 35–57. Springer, Heidelberg (2003). https://doi.org/10.1007/978-3-662-05105-4_2
24. Qiu, W.T.G.: Dense Graph Convolutional Neural Networks on 3D Meshes for 3D Object Segmentation and Classification. arXiv:2106.15778 (2021)
25. Ranjbar, S., Ross Mitchell, J.: An introduction to radiomics: an evolving cornerstone of precision medicine. In: Biomedical Texture Analysis, pp. 223–245. The Elsevier and MICCAI Society Book Series. Academic Press (2017). https://doi.org/10.1016/B978-0-12-812133-7.00008-9

26. Tammemagi, M.C., et al.: Participant selection for lung cancer screening by risk modelling (the Pan-Canadian Early Detection of Lung Cancer [PanCan] study): a single-arm, prospective study. Lancet Oncol. **18**(11), 1523–1531 (2017). https://doi.org/10.1016/S1470-2045(17)30597-1

27. Usman, M., Lee, B., Byon, S.S., Kim, S., Lee, B., Shin, Y.: Volumetric lung nodule segmentation using adaptive ROI with multi-view residual learning. Sci. Rep. **10**(1), 12839 (2020). https://doi.org/10.1038/s41598-020-69817-y

28. Wilson, R., Devaraj, A.: Radiomics of pulmonary nodules and lung cancer. Transl. Lung Cancer Res. **6**(1), 86 (2017)

29. Winter, A., Aberle, D.R., Hsu, W.: External validation and recalibration of the Brock model to predict probability of cancer in pulmonary nodules using NLST data. Thorax **74**(6), 551–563 (2019). https://doi.org/10.1136/thoraxjnl-2018-212413

30. Wu, B., Zhou, Z., Wang, J., Wang, Y.: Joint learning for pulmonary nodule segmentation, attributes and malignancy prediction. In: 2018 IEEE 15th International Symposium on Biomedical Imaging (ISBI 2018), pp. 1109–1113 (2018). https://doi.org/10.1109/ISBI.2018.8363765. ISSN 1945-8452

Self-supervised Approach for a Fully Assistive Esophageal Surveillance: Quality, Anatomy and Neoplasia Guidance

Ziang Xu[1,3] ⓘ, Sharib Ali[2](✉) ⓘ, Numan Celik[1] ⓘ, Adam Bailey[4,5] ⓘ, Barbara Braden[4,5] ⓘ, and Jens Rittscher[1,3,4] ⓘ

[1] Institute of Biomedical Engineering, University of Oxford, Oxford, UK
[2] School of Computing, University of Leeds, Leeds, UK
s.s.ali@leeds.ac.uk
[3] Big Data Institute, Li Ka Shing Centre for Health Information and Discovery, University of Oxford, Oxford, UK
[4] NIHR Oxford Biomedical Research Centre, Oxford, UK
[5] Translational Gastroenterology Unit, John Radcliffe Hospital, Oxford, UK

Abstract. Early pre-cancerous malignant condition such as Barrett's esophagus (BE) and quantification of associated dysplastic changes is critical for early diagnosis and treatment. While endoscopic videos are corrupted with multiple artefacts and procedure require investigating extended areas such as stomach, it is inevitable that there is risk of missing areas that may potentially harbour neoplastic changes and require immediate attention. A complete guidance assisting navigation is thus vital. Visually obvious neoplasia and suspected areas both needs be flagged for biopsies. Due to the thin demarcation between BE and early neoplasia, it is often challenging to identify subtle changes even by experts. We propose a self-supervised learning technique for a fully assistive esophageal endoscopy surveillance system. The self-supervision step allows to learn complex representations using a pretext task of solving a jigsaw puzzle. Here, the idea is to enable network to distinctly learn inconspicuous features that are characteristics of neoplasia and other classes. In order to enable an optimal decision boundary we propose to incorporated angular margin in our fine-tuning process. Our proposed framework showed a boost of 3% on overall accuracy compared to fully supervised approach with similar backbone.

Keywords: Endoscopy · Neoplasia · Deep learning · Classification · Self supervised

1 Introduction

Endoscopy is a routine clinical surveillance technique for diagnosis and minimally invasive treatments. Acid reflux disease, also termed as *heartburn*, is caused due to the acid flowing back into the stomach's food pipe. In gastroesophageal reflux disease, acid flows from the stomach into the food pipe, causing symptoms such

© The Author(s), under exclusive license to Springer Nature Switzerland AG 2022
S. Ali et al. (Eds.): CaPTion 2022, LNCS 13581, pp. 14–23, 2022.
https://doi.org/10.1007/978-3-031-17979-2_2

as heartburn. Some patients with constant reflux over the years can develop a condition called Barrett's esophagus that is associated with a mildly increased risk of adenocarcinoma in the esophagus [7]. During upper endoscopy, a long, flexible tube with a camera at its tip is inserted through the mouth into the esophagus, and the lining of the gastrointestinal tract is observed on a separate monitor. This procedure lasts for about 20 min on average, during which the endoscopists visualize the esophagus, stomach, and upper part of the small bowel [3]. While doing so, the endoscopists try to understand and differentiate between the normal tissue (the normal squamous region in the esophagus), non-dysplastic Barrett's esophagus (NDBE), pre-cancerous neoplastic changes, and even esophageal adenocarcinoma (EAC). Further, endoscopic imaging is challenging due to the presence of a multitude of artifacts such as bubbles, pixel saturation, specularity, bodily fluids, and blur due to organ or camera motion [1]. The impairment of visualization caused by artifacts can be a significant hurdle for endoscopists to miss certain areas that may harbor neoplasia.

In this work, we provide an integrated framework that incorporates the classification of anatomical location and mucosal differentiation (see samples of esophageal endoscopy in Fig. 2), including 1) identification of the frames that have quality issues, 2) classification of video frames with different anatomical area location (e.g., stomach and normal esophagus), and 3) identify precancerous precursors (neoplasia and suspected or early neoplasia). Till date, most developed classification methods focus either on site [3] only or on neoplasia [7] or cancer [15]. To our knowledge, this will be the first work that proposes a comprehensive technique that can enable all anatomical site classification, neoplasia recognition, localization, and quality assurance using self-supervision and decision boundary optimization.

2 Related Work

Various studies have been explored in the literature that aimed at the classification and detection of NDBE, dysplastic BE, and EAC using deep learning methods [7,9,15]. A deep learning-based computer-aided method using ResNet18 was used to classify early neoplasia and NDBE [7]. The classifier showed an accuracy of 89% accuracy with 90% sensitivity and 88% specificity on the test set. Furthermore, the model outperformed the general endoscopists for a classification task by 10–20% in accuracy, sensitivity, and specificity. Liu et al. [15] reported a CNN-based model for the classification of esophageal cancer (EAC) from early neoplasia and normal endoscopic frames. The proposed approach was based on the Inception-ResNet model and comprised of two sub-networks: O-Net for extracting the color and global features and P-Net for extracting the texture and detail features. Their approach achieved an accuracy of 85.83%, a sensitivity of 94.23%, and specificity of 94.67% after the fusion of the two subnetworks. Similarly, a deep learning approach for classification EAC (often obvious lesion) using ResNet and NDBE was developed by Ebigbo et al. [9]. WLI and NBI modalities were used, and the authors reported a sensitivity of 97% and specificity of 88%. A dual-stream network for classification and segmentation of esophageal lesions was proposed by Wu et al. [20]. A large image dataset of 189,436 images was used by Hussein et al. [13] for dysplasia (high-grade dysplasia or EAC) and NDBE classification using ResNet-101. For the anatomical site classification task, He et al. [12] used 3704 images of upper GI endoscopy for training and evaluated the classical models. Their best model using DenseNet-121 provided an overall accuracy of 88%. A similar work to [12] by Park et al. [14] employs several classical CNN-based models to classify anatomical sites that included the esophagus, stomach, and duodenum; for which 11 different annotated sub-classes were present. Their best-performing model was EfficientNet-B1. GoogLeNet was used by Takiyama et al. [18] to classify anatomical sites with a sensitivity of 96.4% and specificity of 98.1%. As one can observe, most previous methods only used fully supervised techniques [7,13,20]. In order to optimally provide a classification of this complex imaging domain, fully supervised methods require a large number of labeled samples incorporating different variability, as they are known for their data voracious nature. Endoscopic appearance and (early) neoplastic changes are often inconspicuous and can add difficulty to the network to understand the representations. Self-supervised methods have allowed us to learn complex and subtle representations in medical images without requiring labels [2]. Azizi et al. [2] used SimCLR [5] for self-supervision on the medical images using multi-instance contrastive learning where two crops of the images from the same patients were drawn and compared. Finally, a fully supervised fine-tuning approach was taken for the final classification of available task-specific loss, showing a performance boost over the classically used fully supervised approach. Chen et al. [4] presented a novel self-supervised learning strategy based on context restoration. The method altered the spatial information of the image by selecting and swapping two small patches in the same image to learn enough useful semantic representations. Three different types of medical imaging datasets were used for the evaluation of classification (on 2D fetal ultrasound images), localization (on abdominal computed tomography images), and segmentation tasks (on brain magnetic resonance images). Previous studies showed a boost in performance on medical imaging datasets when SSL techniques are used [2,4,17]. In this work, we investigate recent state-of-the-art (SOTA) self-supervised methods [5,6,10,16] for developing a comprehensive six class esophageal endoscopy classification task that includes (imaging) artefacts, normal esophagus, non-dysplastic Barrett's (NDBE), neoplasia, suspected or early neoplasia, and stomach. Here, we propose to use the pretext invariant representation learning technique together with an angular margin-based loss function (PIRL-AM) to allow optimal decision boundaries on the learned representations from the pretext task of the SSL network.

3   Method

We propose a self-supervised learning framework that is aimed at solving a jig-saw puzzle and has shown to have invariance properties [16]. Further, to increase inter-class separation and minimize the intra-class distance, we propose to include an angular margin approach [8] in our fine-tuning loss function. The overall framework is shown in Fig. 1.

3.1   Self-supervision Solving Jigsaw Puzzle

Let $\mathcal{D} = \{\mathbf{I}_1, \mathbf{I}_2, ..., \mathbf{I}_N\}$, where N represents image samples present in the dataset. Here, a transformation $\mathcal{T}$ is applied to create and reshuffle jigsaw puzzle that consists of m number of image patches in $\mathcal{D}$, $\mathcal{P} = \{\mathbf{I}_{1t}^1, ..., \mathbf{I}_{1t}^m, ..., \mathbf{I}_{Nt}^1, ..., \mathbf{I}_{Nt}^m\}$ with $\mathcal{T} \in t$. We train a convolutional neural network with free parameters θ that embody feature representations $\phi_\theta(\mathbf{I})$ for a given sample $\mathbf{I}$ and $\phi_\theta(\mathbf{I}_t)$ for patch $\mathbf{I}_t$. For image patches, representations of each patch constituting the image $\mathbf{I}$ is concatenated. A 128-d dimension projections are obtained for each image and patch level representations with functions $f(.)$ and $g(.)$, respectively.

A memory bank $\mathcal{M}$ is used to store all negative and positive samples with the moving average of embedding denoted as $m_\mathbf{I}$ and $m_{\mathbf{I}'}$, respectively. A noise contrastive estimator (NCE) measuring the similarity between two representations and the list of negative samples, say $\mathcal{D}_n$. A unique projection heads, $f(.)$ and $g(.)$, are applied to re-scale the representations to a 128-dimensional feature vector in each case (see Fig. 1). A memory bank is used to store positive and negative sample embedding of a mini-batch B (in our case, $B = 32$). Negative refer to embedding for $I' \neq I$ that is required to compute our contrastive loss function $\mathcal{L}(.,.)$, measuring the similarity between two representations. The list of negative samples, say $\mathcal{D}_n$, grows with the training epochs and are stored in a memory bank $\mathcal{M}$. To compute a noise contrastive estimator (NCE), each positive samples has $|\mathcal{D}_n|$ negative samples and minimizes the loss:

$$\mathcal{L}_{NCE}(\mathbf{I}, \mathbf{I}_t) = -\log[h(f(\phi_\theta(\mathbf{I})), g(\phi_\theta(\mathbf{I}_t)))] - \sum_{I' \in \mathcal{D}_n} \log[1 - h(f(\phi_\theta(\mathbf{I}')), g(\phi_\theta(\mathbf{I}_t)))].$$

$$(1)$$

For our experiments we have used both ResNet50 [11] and a combination of convolutional block attention (CBAM, [19]) with ResNet50 model (ResNet50^{+cbam}) for computing the representation $f(.)$ and $g(.)$. In Eq. (1), $h(.,.)$ is the *cosine* similarity between the representations. The self-supervision task is to learn the representations that enable to correctly classify puzzle pieces $\mathbf{I}_t$ coming from the same image $\mathbf{I}$.

3.2   Fine-Tuning with Angular Margin Loss

To optimize the decision boundary to exploit the learned subtle features that could distinguish between similar classes such as neoplasia and suspected, for example, we introduce an arcFace [8] loss that introduces an angular margin penalty based on softmax. Such a loss function is geometrically interpretable. The weights from the fine-tuning network are first $L2$-normalized, and so are the features from the $\phi_\theta(\mathbf{I})$ such that the learned embeddings are distributed on a unit hyper-sphere. As shown in Fig. 1 (see fine-tuning stage), an angle θ_{y_i} is computed as product between normalised feature embedding and weights where y_i represent logits $i \in [0, 1, 2, ..., N]$. After which a margin 'm' is added to θ_{y_i} updated to θ_{y_j} $(\theta_{y_i} + m)$. Doing so increases the decision boundary between classes pushing interclass samples and pulling intra-class samples. Finally, a scaling factor 's' is used to

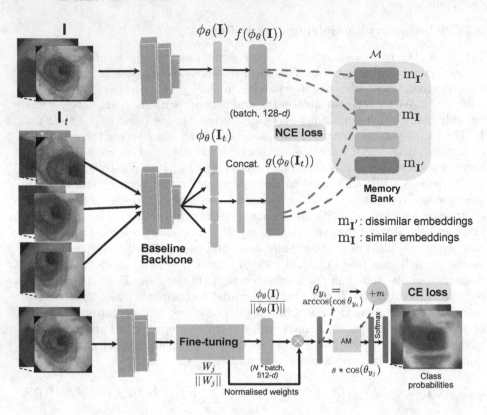

Fig. 1. Self-supervised approach with additive angular margin (AM) as fine-tuning loss for improved fully assistive esophageal surveillance. Here, we use two identical classification networks to compute both image-level and patch-level representation learning. Patch level representations indicate the transformed images that enable pretext invariant representation learning [16]. To do so, a memory bank is used that contains the moving average of representations for all images in the dataset. Here, **I** represents an image sample while **I**$_t$ is a transformed puzzle of that image, and **I**$'$ represents a negative sample. Our fine-tuning approach adds an angular margin 'm' between the feature representations and normalised weights ($\theta_{y_j} = \theta_{y_i} + m$) in order to maximise class separability [8]. Finally, a scaling factor 's' is added to rescale the output.

scale these projected features for subsequent softmax-based class probability predictions. It is to be noted that the angle m is placed inside the $\cos\theta$ function such that $\cos(\theta + m)$ is smaller than $\cos\theta$ in the range of $\theta \in [0, \pi - m]$. The angular margin loss is given by:

$$\mathcal{L}_{AM} = -log \frac{e^{s\cos(\theta_{y_i}+m)}}{e^{s\cos(\theta_{y_i}+m)} + \sum_{j=1, j\neq y_i}^{N} e^{s\cos(\theta_j)}}, \tag{2}$$

where θ_j is the angle between the embedding feature x_i of the i-th sample and the j-th ground truth (target) weight W_j which behaves as a center for each class, here, y_i represents the corresponding class label of x_i, N is the total class number, and s and m are the parameter of feature re-scale and angular margin, respectively. Cross entropy loss is then minimized between ground truth and label probabilities (see Fig. 1, standard CE loss). It has been empirically shown [8] that setting $m = 0.5$ and $s = 30$ provides optimal results.

4   Experiments and Results

4.1   Implementation Details

PyTorch-based[1] implementations were used for all methods. All images in our experiments were resized to 224×224 pixels. In pretext tasks of self-supervised learning, an SGD optimizer with a weight decay of $4e^{-4}$, the momentum of 0.9, and a learning rate of $1e^{-4}$ were used for all experiments. Three thousand iterations with a batch size of 64 were used to train pretext tasks.

For the downstream classification task, finetuning with a batch size of 64, the Adam optimizer and the learning rate of $1e^{-4}$ were used for the 6-classes classification task. We also use a learning rate decay strategy with a learning rate decay of 0.9 times per 10 epochs. All experiments were implemented on a server deployed with an NVIDIA Quadro RTX 6000 card.

4.2   Data Collection and Evaluation Metrics

Upper GI endoscopic images were obtained from the University of Oxford, John Radcliffe Hospital. There were 2103 images collected from over 132 unique patient videos. The acquired images consisted of two different modalities, white light imaging (WLI) and narrow-band imaging (NBI), as it occurs in routine endoscopy, and manually classified into six categories by two experienced gastroenterologists. No separation between modalities was done to mimic real-world clinical endoscopy (Table 1).

Table 1. Number of images in each training and test datasets.

Categories	Train set	Test set
Artefact	290	84
Neoplasia	270	100
NDBE	164	72
Normal	210	66
Stomach	285	86
Suspected	124	48

---

[1] https://pytorch.org.

Figure 2 shows representative images of 6 categories as follows: 1) artefact, 2) non-dysplastic Barrett's esophagus (NDBE), 3) neoplasia, 4) normal, 5) suspected, and 6) stomach. The images with poor quality, including excessive mucus, food residue, and active bleeding after endoscopic resection, were removed. In our study, we defined low-grade dysplasia as a suspected category to avoid any confusion in deciding on the precancerous lesion in the endoscopic image, while both high-grade and tumors were included in the neoplasia category.

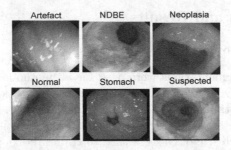

Fig. 2. Representative images of different classes during an esophageal endoscopic surveillance

We have used standard F1-score $(= \frac{2tp}{2tp+fp+fn}$, tp: true positive, fp: false positive), specificity $(= \frac{tp}{tp+fn})$, sensitivity $(= \frac{tn}{tn+fp})$ and top-k accuracy (percentage of samples predicted correctly) for our classification task for esophageal endoscopy.

Table 2. Quantitative results on various self-supervised SOTA, fully-supervised baseline and our proposed approach

Method	Model	Top 1	Top 2	F1-score	Specificity	Sensitivity
Baseline	ResNet50	80.48%	91.07%	79.59%	95.81%	78.04%
	ResNet50 +CBAM	81.14%	91.74%	80.36%	96.33%	78.69%
SimCLR [5]	ResNet50 +CBAM	80.26%	90.36%	79.17%	95.28%	75.32%
SimCLR + DCL [6]	ResNet50 +CBAM	82.23%	92.13%	81.22%	97.36%	77.01%
MOCO + CLD [10]	ResNet50 +CBAM	82.89%	94.89%	82.31%	97.91%	80.83%
PIRL [16]	ResNet50 +CBAM	83.33%	93.29%	83.02%	98.35%	82.50%
PIRL-AM (proposed)	ResNet50 +CBAM	84.09%	93.44%	83.37%	98.02%	82.73%

4.3   Comparison with SOTA Methods

From Table 2, baseline networks ResNet50 and ResNet50 with convolutional block attention module (CBAM) respectively obtained 80.48% and 81.14% on top-1 accuracy. Compared to the ResNet50, ResNet50 with CBAM improved nearly 1%

on top-1 accuracy, F1-score, specificity, and sensitivity, respectively. Therefore, for each experiment with the SOTA SSL methods and the proposed method, we have used ResNet50+CBAM as the backbone architecture. The proposed method (PIRL-AM) achieved 84.09% and 83.37% on top-1 accuracy and F1-score, which got 2.95% and 3.01% improvement compared to the fully supervised learning-based baseline model. Our experiments on all the existing SOTA SSL methods

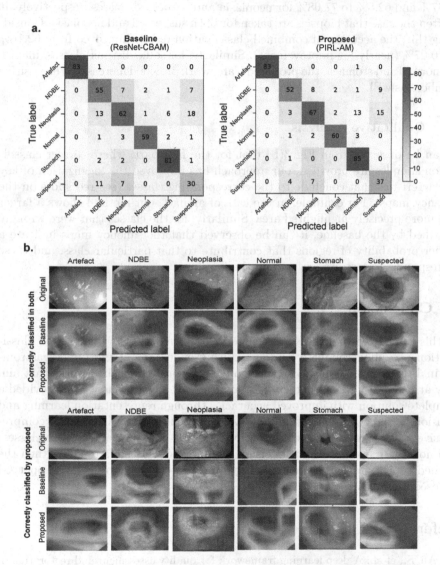

Fig. 3. a) Confusion matrix for both baseline and proposed network with the same backbone (ResNet+CBAM). b) Qualitative results on the baseline and proposed self-supervised approach showing correctly classified samples by both (top) and miss classified by baseline but correctly classified by our approach. Saliency maps are provided for each sample.

showed that our proposed method has improved performance. Top 1 accuracies for SimCLR presented 80.26% (3.83% below ours), SimCLR+DCL presented 82.23% (1.86% below ours), MOCO+CLD with 82.89% (1.2% below ours) and for PIRL 83.33% (0.76% below ours). Also, from the provided confusion matrix in Fig. 3, we can observe a substantial improvement between neoplasia and suspected classes (tp of 62 to 67 and tp of 30 to 37, respectively). The accuracy increased from 62% to 67% and 62.5% to 77.08% for neoplasia and suspected classes, respectively. It is often the case that biopsies are taken for both suspected and neoplasia. Considering this, the accuracy of combined classification results improved from 62.16% to 70.27% (nearly 8% improvement). Similarly, for some distinct classes, including normal and stomach, the proposed framework provided increased true positive numbers as well.

4.4 Qualitative Analysis

It can be observed from Fig. 3(b) that for the first part, where correct classification samples are provided, our approach has improved the localization of key characteristics that conclude to the corresponding classes (see red spots on the saliency maps). For example, in the case of artifact, our method shows a larger and more precisely highlighted area. Similarly, for the classes that were wrongly classified by the baseline, it can be observed that the saliency maps indicate a higher probability of regions that contribute to that particular class, and these corresponding areas are larger than the baseline.

5 Conclusion

In this paper, we proposed a self-supervised framework for multi-class classification providing complete guidance during endoscopy of the upper gastrointestinal tract of the upper GI tract. Unlike previously published works that aim only at neoplastic lesions detection or anatomy or artifact classes, we provided a complete solution with improved accuracies through representation learning and decision boundary optimization. Our work is the first that includes the comprehensive, diverse multi-class classification (quality, anatomy, and disease classes) and novelty in optimizing the decision boundary of classifiers to strengthen the learned representations demonstrating improved classification of the upper GI endoscopy frames.

References

1. Ali, S., et al.: A deep learning framework for quality assessment and restoration in video endoscopy. Med. Image Anal. **68**, 101900 (2021)
2. Azizi, S., et al.: Big self-supervised models advance medical image classification. In: 2021 IEEE/CVF International Conference on Computer Vision (ICCV), pp. 3458–3468 (2021)

3. Chang, Y.Y., et al.: Development and validation of a deep learning-based algorithm for colonoscopy quality assessment. Surg. Endosc. (2022)
4. Chen, L., Bentley, P., Mori, K., Misawa, K., Fujiwara, M., Rueckert, D.: Self-supervised learning for medical image analysis using image context restoration. Med. Image Anal. **58**, 101539 (2019)
5. Chen, T., Kornblith, S., Norouzi, M., Hinton, G.: A simple framework for contrastive learning of visual representations. In: International Conference on Machine Learning, pp. 1597–1607. PMLR (2020)
6. Chuang, C.Y., Robinson, J., Lin, Y.C., Torralba, A., Jegelka, S.: Debiased contrastive learning. Adv. Neural. Inf. Process. Syst. **33**, 8765–8775 (2020)
7. de Groof, A.J., et al.: Deep-learning system detects neoplasia in patients with Barrett's esophagus with higher accuracy than endoscopists in a multistep training and validation study with benchmarking. Gastroenterology **158**(4), 915-929.e4 (2020)
8. Deng, J., Guo, J., Xue, N., Zafeiriou, S.: Arcface: additive angular margin loss for deep face recognition. In: Proceedings of the IEEE/CVF Conference on Computer Vision and Pattern Recognition, pp. 4690–4699 (2019)
9. Ebigbo, A., et al.: Computer-aided diagnosis using deep learning in the evaluation of early oesophageal adenocarcinoma. Gut **68**(7), 1143–1145 (2019)
10. He, K., Fan, H., Wu, Y., Xie, S., Girshick, R.: Momentum contrast for unsupervised visual representation learning. In: Proceedings of the IEEE/CVF Conference on Computer Vision and Pattern Recognition, pp. 9729–9738 (2020)
11. He, K., Zhang, X., Ren, S., Sun, J.: Deep residual learning for image recognition. In: Proceedings of the IEEE Conference on Computer Vision and Pattern Recognition, pp. 770–778 (2016)
12. He, Q., et al.: Deep learning-based anatomical site classification for upper gastrointestinal endoscopy. Int. J. Comput. Assist. Radiol. Surg. **15**(7), 1085–1094 (2020). https://doi.org/10.1007/s11548-020-02148-5
13. Hussein, M., et al.: Deep neural network for the detection of early neoplasia in Barrett's oesophagus. Gastrointest. Endosc. **91**(6), AB250 (2020)
14. Jung-Whan, P., Yoon, K., Woo-Jin, K., Seung-Joo, N.: Automatic anatomical classification model of esophagogastroduodenoscopy images using deep convolutional neural networks for guiding endoscopic photodocumentation. J. Korea Soc. Comput. Inf. **26**(3), 19–28 (2021)
15. Liu, G., et al.: Automatic classification of esophageal lesions in endoscopic images using a convolutional neural network. Ann. Transl. Med. **8**(7) (2020)
16. Misra, I., Maaten, L.V.D.: Self-supervised learning of pretext-invariant representations. In: Proceedings of the IEEE/CVF Conference on Computer Vision and Pattern Recognition, pp. 6707–6717 (2020)
17. Ouyang, C., Biffi, C., Chen, C., Kart, T., Qiu, H., Rueckert, D.: Self-supervised learning for few-shot medical image segmentation. IEEE Trans. Med. Imaging (2022). https://doi.org/10.1109/TMI.2022.3150682
18. Takiyama, H., et al.: Automatic anatomical classification of esophagogastroduodenoscopy images using deep convolutional neural networks. Sci. Rep. **8**(1), 1–8 (2018)
19. Woo, S., Park, J., Lee, J.Y., Kweon, I.S.: CBAM: convolutional block attention module. In: Proceedings of the European Conference on Computer Vision (ECCV), pp. 3–19 (2018)
20. Wu, Z., Ge, R., Wen, M., Liu, G., et al.: ELNet: automatic classification and segmentation for esophageal lesions using convolutional neural network. Med. Image Anal. **67**, 101838 (2021)

Multi-scale Deformable Transformer for the Classification of Gastric Glands: The IMGL Dataset

Panagiotis Barmpoutis[1,2]($\boxtimes$), Jing Yuan[3], William Waddingham[2], Christopher Ross[2], Kayhanian Hamzeh[2], Tania Stathaki[3], Daniel C. Alexander[1], and Marnix Jansen[2]

[1] Department of Computer Science, University College London, London, UK
p.barmpoutis@ucl.ac.uk
[2] Department of Pathology, University College London, London, UK
[3] Department of Electrical and Electronic Engineering, Imperial College London, London, UK

Abstract. Gastric cancer is one of the most common cancers and a leading cause of cancer-related death worldwide. Among the risk factors of gastric cancer, the gastric intestinal metaplasia (IM) has been found to increase the risk of gastric cancer and is considered as one of the precancerous lesions. Therefore, early detection of IM could allow risk stratification regarding the possibility of progression to cancer. To this end, accurate classification of gastric glands from the histological images plays an important role in the diagnostic confirmation of IM. To date, although many gland segmentation approaches have been proposed, no general model has been proposed for the identification of IM glands. Thus, in this paper, we propose a model for gastric glands' classification. More specifically, we propose a multi-scale deformable transformer-based network for glands' classification into normal and IM gastric glands. To evaluate the efficiency of the proposed methodology we created the IMGL dataset consisting of 1000 gland images, including both intestinal metaplasia and normal cases received from 20 Whole Slide Images (WSI). The results showed that the proposed approach achieves an F1 score equal to 0.94, showing great potential for the gastric glands' classification.

Keywords: Medical image classification · Vision transformers · Gastric cancer · Intestinal metaplasia

1 Introduction

Gastric cancer is one of the most frequent causes of cancer-related deaths worldwide. As reported by the WHO in 2020 [1], it is the sixth most frequent type of cancer and it is the fourth leading cause of cancer-related deaths mainly due to its often-late stage of diagnosis [2]. The risk factors of gastric cancer include Helicobacter pylori infection, salt intake, tobacco smoking, alcohol consumption, family history of gastric cancer, gastric atrophy and intestinal metaplasia (IM) [2, 3]. More specifically, the IM of the mucosa of the stomach is a major precursor lesion that is associated with an increased risk of dysplasia and cancer [4, 5]. For this reason, early and effective diagnosis of IM is a

© The Author(s), under exclusive license to Springer Nature Switzerland AG 2022
S. Ali et al. (Eds.): CaPTion 2022, LNCS 13581, pp. 24–33, 2022.
https://doi.org/10.1007/978-3-031-17979-2_3

crucial step to prevent gastric cancer. In the IM, the native gastric glands are replaced by metaplastic glands and gastric mucinous epithelial cells are replaced by goblet cells, enterocytes and colonocytes. Widely used diagnostic methods for IM include endoscopic and histological diagnosis. Endoscopic diagnosis of severe cases of IM is effortless, but there are difficulties in making the diagnosis of mild IM cases. Therefore, a biopsy of suspected cases of IM is suggested. Then, based on the Sydney protocol [6], IM is histologically diagnosed using hematoxylin and eosin (H&E) stain.

However, the visual assessment of glands by histopathologists is a laborious and time-consuming task [7]. Thus, the automated precise segmentation and classification of glands from the histological images plays an important role in the morphological analysis of glands, which is a crucial criterion for effective IM detection and management. Numerous methods have been proposed in literature for gland segmentation. However, to date, no generally applicable digital pathology approach has been proposed and applied for gastric glands' classification and more specifically for the identification and analysis of gastric intestinal metaplastic glands. Towards this end, in this paper, we propose a new methodology for gastric glands' classification based on H&E-stained images. More specifically, this paper makes the following contributions:

- We propose the IMGL-VTNet (Intestinal Metaplasia gastric GLands-Vision Transformer Net) that integrates a multi-scale deformable transformer model and a focal loss function for the gastric glands' classification.
- We publish the annotated IMGL dataset (Intestinal Metaplasia gastric GLands) that consists of normal and IM cases that we used for the training and testing of the proposed model. As a small number of research studies of gastric tissues use public data [8], we anticipate this dataset will provide the foundation for advanced studies of IM gastric glands and biopsies.

The rest of this paper is organized as follows: First, details of the proposed methodology are presented, followed by experimental results using the IMGL dataset. Finally, some conclusions are drawn and future extensions are discussed.

2   Related Works

The digital medical image classification field receives growing attention and has become increasingly popular. Thus, various techniques and methods, based on either hand-crafted or deep learning features, have been developed for histopathological image classification tasks. Hand-crafted developed classification approaches for digital pathology tasks are based on grayscale density, color, texture and shape information [9–11]. After the extraction of low-level or mid-level set of features, post-processing methods such as dimensionality reduction and a classifier are usually used aiming to assign a classification label to each image [12]. On the other hand, more sophisticated classification methods such as deep-learning techniques [13] and higher-order dynamical systems [14, 15] have been developed aiming to address medical and histopathological image classification problems by extracting high-level features and knowledge directly from the data.

More recently, vision transformers inspired by the deep learning model that developed for the Natural Language Processing (NLP) [16] have been utilized for medical image segmentation [17], classification [18] and various computer vision tasks. Vision transformers apply attention mechanism to quantify pairwise long-range entity interactions [19]. These can be used as the form of self-attention layers or encoder-decoder pairs. More specifically, an adaptation of the BoTNet [19] has been proposed for image classification replacing the spatial convolutional layers with multi-head self-attention (MHSA) layers in the last stage of ResNet. In contrast, i-ViT [20] uses the transformer encoder to extract and aggregate features of instance patches for the papillary renal cell carcinoma subtyping task. The deformable DETR [21] is a fast-converging and memory-saving vision transformer with six encoder-decoder pairs, which facilitates high resolution feature maps from multiple scales. Owing to the efficiency, DT-MIL [22] applies it to high-level bag representation for multi-instance learning on histopathological images. Inspired by the deformable DETR, we propose a model that adopts a vision transformer in the glands' classification task aiming to exploit the local and global visual dependencies utilizing multi-scale deformable self-attention and a novel scale-aware feature extraction module.

3   Materials and Methods

The framework of the proposed methodology for the gastric glands' classification into normal and IM cases is shown in Fig. 1. Initially, the manually annotated IMGL dataset based on 20 WSI was created. Then, the segmented glands were fed to the proposed IMGL-VTNet for the classification of gastric glands.

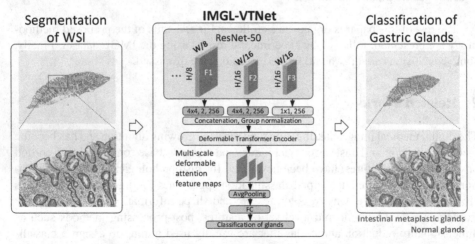

Fig. 1. The proposed methodology. The IMGL-VTNet takes the advantage of the deformable transformer encoder to extract multi-scale features.

3.1 IMGL Dataset Description

To evaluate the efficiency of the proposed methodology we created the IMGL dataset consisting of gastric glands (Fig. 2). More specifically, the dataset includes 500 normal and 500 IM gastric glands. Gastric tissues were collected at University College London Hospital NHS trust, with ethical approval (research ethics committee (REC) reference: 15/YH/0311, & 19/LO/0089) with informed consent taken for prospective tissue collection. The tissues underwent routine H&E staining. For the evaluation of the IMGL-VTNet model we used five-fold cross validation selecting 800 gland images for the training and 200 images for the testing. It is worth mentioning that, as our aim is to develop a methodology for the early detection and diagnosis of IM to prevent gastric cancer, in this dataset we included mild and moderate IM cases. The dataset is available at the following link: https://doi.org/10.5281/zenodo.6908133.

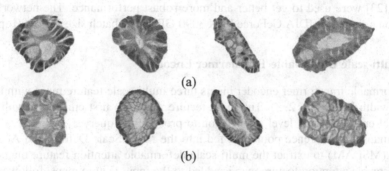

(a)

(b)

Fig. 2. Dataset images including (a) IM gastric glands and (b) normal gastric glands.

3.2 The Proposed IMGL-VTNet Architecture

The proposed model uses the ResNet-50 as the backbone, followed by the deformable transformer encoder-based feature extraction module. More specifically, in order to extract higher-level semantic information preserving the resolution, the stride and dilation of the last stage of the backbone are set as 1 and 2 respectively. Then, feature maps F_1 and F_2 were upsampled by two, while F_3 was encoded with a convolutional layer. Different kernel sizes were applied to each feature map as shown in Fig. 1. Then, the multi-scale feature maps were concatenated, and group normalized and were fed into a deformable transformer encoder for the extraction of multi-scale features and the exploitation of local and global dependencies. Moreover, the extracted multi-level features were used, and an average pooling was considered followed by a fully connected layer for the classification of gastric glands into normal and IM.

To further enhance the model performance, a modulation term was applied to the binary Cross-Entropy loss function. The resulted focal loss [24] focuses on a set of hard examples improving the precision for these cases. More specifically, we defined the following loss function FL_i for the i-th image:

$$FL_i = w_{focal} \cdot Loss \qquad (1)$$

$$w_{focal} = \begin{cases} (1-s)^\gamma & p=1 \\ s^\gamma & p=0 \end{cases} \tag{2}$$

$$Loss = p\log(s) + (1-p)\log(1-s) \tag{3}$$

where p is the ground truth (0 or 1) that represents the two categories (normal and IM), s is the predicted score and γ is the predesigned hyperparameter (we set $\gamma = 2$). It is worth mentioning that as the two categories of the IMGL dataset have the same number of training images, no additional balance was needed.

The input images were first resized and padded to the fixed shape of (224, 224). In addition, an augmentation method was utilized to further increase the variability of the training dataset and to avoid overfitting of the network. In particular, we included translation, rotation and flipping transformations. The Adam optimizer and mean teacher method [23] were used to get better and more robust performance. The network was trained on a single NVIDIA GeForce RTX 3090 GPU with batch size 16 for 80 epochs.

3.3 Multi-scale Deformable Transformer Encoder

The deformable transformer encoder inputs three multi-scale feature maps with height H_l and width $W_l(l = 1, 2, 3)$. The input feature maps are first embedded with fixed positional encodings and level information to produce the query z_q. The query, input feature maps and reference points are fed into the Multi-Scale Deformable Attention Module (MSDAM) to extract the multi-scale deformable attention feature map. Then the deformable attention feature map is added to the input feature maps, followed by a Feed-Forward Network (FFN).

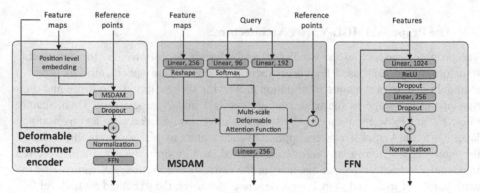

Fig. 3. Deformable transformer encoder consisting of a Multi-scale Deformable Attention Module (MSDAM) and a Feed-Forward Network (FFN).

In the MSDAM, value, weight and location tensors are first computed and applied to the multi-scale deformable attention function to produce the multi-scale deformable attention feature map z_o via weighted average. As shown in Fig. 3, the value tensor v is produced by embedding the input features via a linear layer. The weight W and sampling offsets Δp are produced by embedding the query via two linear layers respectively. The weight is further normalized by a softmax operator along the scale and sampling point dimensions. The sampling location is the element-wise addition of sampling offset Δp and the reference points p. More specifically, the q-th element of the separate deformable attention feature $z' \in \mathbb{R}^{N_q \times c_v} (N_q = \sum_{l=1}^{3} H_l W_l)$ at a single head is expressed as follows:

$$z'_q = \sum_p^{N_p} \sum_{l=1}^{3} W_{plhq} v_{p_{ql} + \Delta p_{qhlp}} \tag{4}$$

where q, h and p denote the elements of the deformable attention feature z_o, the attention head, and the sampling offsets respectively. W_{plhq} is an entity of $W \in \mathbb{R}^{N_q \times N_h \times 3 \times N_p}$. Furthermore, p_{ql} and Δp_{qhlp} denote the position of a reference point and one of the N_p corresponding sampling offset of $p \in \mathbb{R}^{N_q \times 3 \times 2}$ and $\Delta p \in \mathbb{R}^{N_q \times N_h \times 3 \times N_p \times 2}$ respectively. The number of sampling offsets and attention head are set as $N_p = 4$ and $N_h = 8$. The separate deformable attention features from 8 attention heads are projected to the q-th element of the overall output deformable attention feature z_o by a linear layer:

$$z_{oq} = \sum_{h=1}^{N_h} W'_h z'_{qh} \tag{5}$$

where $W'_h \in \mathbb{R}^{c \times c_v}$ and vector $z'_{qh} \in \mathbb{R}^{c_v}$ denote the learnable weight and the q-th separate deformable attention feature z'_q obtained at h-th attention head.

4 Experimental Results

In this section, we present an evaluation analysis of the proposed gastric gland classification model as well as the efficiency of multi-scale feature maps for glands' classification. The goal of this experimental evaluation is threefold. Initially, we compared the efficiency of gastric glands' classification, using the IMGL dataset and widely used and state-of-the-art approaches. Secondly, we explored the efficiency of multi-scale deformable attention feature maps extracted from the deformable transformer encoder. Finally, to demonstrate the generality of our model, we applied the proposed method to the pedestrian detection task.

To evaluate the performance of the proposed model, we randomly partitioned the dataset into fivefold training and testing sets and we used precision, recall and F1-score.

4.1 A Comparison of State-of-the-Art Methods: IMGL Dataset

In this section, using the IMGL dataset we aim to present a comparison of the proposed methodology against a number of classification approaches. More specifically, in Table 1, we present the evaluation results of the IMGL-VTNet model in comparison to seven classification models. For the comparison, we consider the most widely used models including the state-of-the-art BoTNet-50 [19] architecture that achieves a strong performance on the ImageNet benchmark and has been applied on various tasks.

The results show that the proposed glands' classification approach achieves precision equal to 0.95 and recall equal to 0.94. Moreover, the proposed model achieves F1 score equal to 0.94. The proposed model achieves an F1 score improvement of 0.05 compared to the widely used ResNet-50. Furthermore, the integration of a Multi Head Self-Attention block in ResNet-50 improves the F1 score 0.02. Thus, the proposed model improves the F1 score by 0.03 compared to BotNet-50.

Further experimental results in 39 unannotated WSI (Fig. 4) show that the IMGL-VTNet is robust under various cases. It is worth mentioning that normal cases (Fig. 4a) include only normal glands, while IM WSI (Fig. 4b) include both normal and IM glands. Thus, as it is shown in Fig. 4a, only a very small number of glands are misclassified as IM glands. Further analyses of the unannotated normal cases show that less than 3% of the glands have been misclassified.

Table 1. A comparison of glands' classification using different models.

Method	Precision	Recall	F1 score
ResNet-18	0.92 ± 0.04	0.84 ± 0.03	0.88 ± 0.03
ResNet-50	0.91 ± 0.03	0.86 ± 0.03	0.89 ± 0.03
ResNet-101	0.91 ± 0.03	0.82 ± 0.03	0.86 ± 0.03
VGG-19	0.89 ± 0.03	0.89 ± 0.02	0.88 ± 0.02
Inception-V3	0.91 ± 0.04	0.81 ± 0.03	0.86 ± 0.04
Xception	0.82 ± 0.05	0.78 ± 0.04	0.79 ± 0.04
BotNet-50	0.92 ± 0.03	0.90 ± 0.02	0.91 ± 0.02
IMGL-VTNet (proposed)	**0.95 ± 0.03**	**0.94 ± 0.02**	**0.94 ± 0.03**

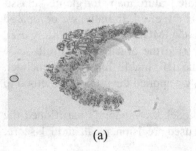   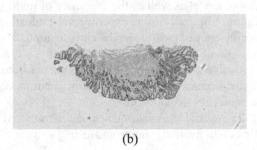

(a)   (b)

Fig. 4. Glands' classification results of IMGL-VTNet model on two sample WSI: a) normal case, b) IM case. Blue color denotes the glands that have been detected as normal and red color denotes the glands that have been detected as IM glands. (Color figure online).

4.2 Feature Map Scales Analysis

Finally, we internally investigated the efficiency of multi-scale deformable attention feature maps for glands' classification. Thus, we compared the individual use of a single deformable attention feature map instead of the multiple deformable attention feature maps. More specifically, the use of multi-scale feature maps slightly improves the

F1-score by 0.01 (Table 2). The results show that higher-level features achieve better precision while lower-level features achieve better recall score.

Table 2. A comparison of glands' classification efficiency using multi-scale deformable attention feature maps.

Feature map scale	Precision	Recall	F1 score
$W/16 \times H/16$	0.91 ± 0.03	0.96 ± 0.02	0.93 ± 0.02
$W/8 \times H/8$	0.96 ± 0.02	0.92 ± 0.02	0.93 ± 0.01
$W/4 \times H/4$	0.95 ± 0.03	0.93 ± 0.02	0.93 ± 0.02
Multi-scale (IMGL-VTNet)	**0.95 ± 0.03**	**0.94 ± 0.02**	**0.94 ± 0.03**

4.3 Application of the Proposed Model to Pedestrian Detection

Finally, to demonstrate the generality of our model, we applied the VTNet to the pedestrian detection task. The average pooling and fully connected layers were replaced by two parallel branches predicting the confidence score and the corresponding bounding boxes respectively. For the evaluation, the Caltech pedestrian dataset was used [25], which contains approximately 2.5 h of video. The performance was assessed in terms of log-average miss rate over false positives per image denoted as MR^{-2}. Based on the same training and testing protocol, the proposed VTNet outperforms other state-of-the-art pedestrian detectors by reducing the miss rate to 4.1% (Table 3).

Table 3. A comparison of the proposed architecture with five state-of-the-art detectors on the Caltech pedestrian dataset.

Method	MR^{-2} (%)
Faster R-CNN [26]	8.7
ALFNet [27]	8.1
RepLoss [28]	5.0
CSP [29]	4.5
Proposed	**4.1**

5 Conclusion

Multiple risk factors and a multistep process have been associated with gastric carcinogenesis. Among these factors, gastric IM of the mucosa has been recognized as a high-risk precancerous lesion for dysplasia and gastric cancer. However, as the manual assessment of biopsies by histopathologists based on the Sydney System is a laborious and time-consuming task, the accurate detection of IM gastric glands necessitates the adoption of artificial intelligence methods. Thus, in this paper we presented a methodology for the automated classification of gastric glands into normal and IM glands. The proposed IMGL-VTNet model for gastric glands' classification achieves an F1 score equal to 0.94. The results suggest that the proposed methodology obtains promising classification performance on the IMGL dataset. However, limitations of this study include the lack of an end-to-end gland segmentation and classification model that could be adopted on a widespread basis in routine histopathological practice.

Acknowledgments. The EPSRC and CRUK support this work through joint funding in grant number NS/A000069/1.

References

1. WHO: Cancerm. https://www.who.int/news-room/fact-sheets/detail/cancer. Accessed 24 July 2022
2. Waddingham, W., et al.: Recent advances in the detection and management of early gastric cancer and its precursors. Frontline Gastroenterol. **12**(4), 322–331 (2021)
3. Jencks, D.S., Adam, J.D., Borum, M.L., Koh, J.M., Stephen, S., Doman, D.B.: Overview of current concepts in gastric intestinal metaplasia and gastric cancer. Gastroenterol. Hepatol. **14**(2), 92 (2018)
4. Busuttil, R.A., Boussioutas, A.: Intestinal metaplasia: a premalignant lesion involved in gastric carcinogenesis. J. Gastroenterol. Hepatol. **24**(2), 193–201 (2009)
5. Pellegrino, C., et al.: From Sidney to OLGA: an overview of atrophic gastritis. Acta Bio Medica Atenei Parmensis. **89**(Suppl 8), 93 (2018)
6. Dixon, M.F., Genta, R.M., Yardley, J.H., Correa, P.: Classification and grading of gastritis: the updated Sydney system. Am. J. Surg. Pathol. **20**(10), 1161–1181 (1996)
7. Sirinukunwattana, K., et al.: Gland segmentation in colon histology images: the glas challenge contest. Med. Image Anal. **1**(35), 489–502 (2017)
8. Gonçalves, W.G., Dos Santos, M.H., Lobato, F.M., Ribeiro-dos-Santos, Â., de Araújo, G.S.: Deep learning in gastric tissue diseases: a systematic review. BMJ Open Gastroenterol. **7**(1), e000371 (2020)
9. Dimitropoulos, K., Barmpoutis, P., Koletsa, T., Kostopoulos, I., Grammalidis, N.: Automated detection and classification of nuclei in pax5 and H&E-stained tissue sections of follicular lymphoma. SIViP **11**(1), 145–153 (2017)
10. Korkmaz, S.A., Binol, H.: Classification of molecular structure images by using ANN, RF, LBP, HOG, and size reduction methods for early stomach cancer detection. J. Mol. Struct. **15**(1156), 255–263 (2018)
11. Barmpoutis, P., Kayhanian, H., Waddingham, W., Alexander, D.C., Jansen, M.: Three-dimensional tumour microenvironment reconstruction and tumour-immune interactions' analysis. In: Proceedings of the IEEE DICTA, pp. 01–06 (2021)
12. England, J.R., Cheng, P.M.: Artificial intelligence for medical image analysis: a guide for authors and reviewers. Am. J. Roentgenol. **212**(3), 513–519 (2019)
13. Barmpoutis, P., et al.: Tertiary lymphoid structures (TLS) identification and density assessment on H&E-stained digital slides of lung cancer. PLoS ONE **16**(9), e0256907 (2021)

14. Barmpoutis, P., Dimitropoulos, K., Apostolidis, A., Grammalidis, N.: Multi-lead ECG signal analysis for myocardial infarction detection and localization through the mapping of Grassmannian and Euclidean features into a common Hilbert space. Biomed. Signal Process. Control 1(52), 111–119 (2019)

15. Dimitropoulos, K., Barmpoutis, P., Zioga, C., Kamas, A., Patsiaoura, K., Grammalidis, N.: Grading of invasive breast carcinoma through Grassmannian VLAD encoding. PLoS ONE 12(9), e0185110 (2017)

16. Devlin, J., Chang, M.W., Lee, K., Toutanova, K.B.: Pre-training of deep bidirectional transformers for language understanding. arXiv:1810.04805 (2018)

17. Hatamizadeh, A., et al.: Unetr: Transformers for 3d medical image segmentation. In: Proceedings of the IEEE/CVF WACV 2022, pp. 574–584 (2022)

18. Dai, Y., Gao, Y., Liu, F.: Transmed: transformers advance multi-modal medical image classification. Diagnostics. 11(8), 1384 (2021)

19. Srinivas, A., Lin, T.Y., Parmar, N., Shlens, J., Abbeel, P., Vaswani, A.: Bottleneck transformers for visual recognition. In: Proceedings of the IEEE/CVF Conference on Computer Vision and Pattern Recognition 2021, pp. 16519–16529 (2021)

20. Gao, Z., et al.: Instance-based vision transformer for subtyping of papillary renal cell carcinoma in histopathological image. In: de Bruijne, M., et al. (eds.) MICCAI 2021. LNCS, vol. 12908, pp. 299–308. Springer, Cham (2021). https://doi.org/10.1007/978-3-030-87237-3_29

21. Zhu, X., Su, W., Lu, L., Li, B., Wang, X., Dai, J.: Deformable detr: Deformable transformers for end-to-end object detection. arXiv:2010.04159 (2020)

22. Li, H., et al.: DT-MIL: Deformable transformer for multi-instance learning on histopathological image. In: Proceedings of the MICCAI 2021, pp. 206–216 (2021)

23. Tarvainen, A., Valpola, H.: Mean teachers are better role models: weight-averaged consistency targets improve semi-supervised deep learning results. Adv. Neural Inf. Process. Syst. 2017, 30 (2017)

24. Lin, T.Y., Goyal, P., Girshick, R., He, K., Dollár, P.: Focal loss for dense object detection. In: Proceedings of the IEEE ICCV 2017, pp. 2980–2988 (2017)

25. Dollar, P., Wojek, C., Schiele, B., Perona, P.: Pedestrian detection: an evaluation of the state of the art. IEEE Trans. Pattern Anal. Mach. Intell. 34(4), 743–761 (2011)

26. Ren, S., He, K., Girshick, R., Sun, J.: Faster R-CNN: towards real-time object detection with region proposal networks. Adv. Neural Inf. Process. Syst. 2015, 28 (2015)

27. Liu, W., Liao, S., Hu, W., Liang, X., Chen, X.: Learning efficient single-stage pedestrian detectors by asymptotic localization fitting. In: Proceedings of the ECCV 2018, pp. 618–634 (2018)

28. Wang, X., Xiao, T., Jiang, Y., Shao, S., Sun, J., Shen, C.: Repulsion loss: Detecting pedestrians in a crowd. In: Proceedings of the IEEE Conference on Computer Vision and Pattern Recognition 2018, pp. 7774–7783 (2018)

29. Liu, W, Liao, S., Ren, W., Hu, W., Yu, Y.: High-level semantic feature detection: a new perspective for pedestrian detection. In: Proceedings of the IEEE/CVF Conference on Computer Vision and Pattern Recognition 2019, pp. 5187–5196 (2019)

Parallel Classification of Cells in Thinprep Cytology Test Image for Cervical Cancer Screening

Maosong Cao[1], Xin Zhang[2], Xiangshan Fan[3], Lichi Zhang[2],
and Qian Wang[1(✉)]

[1] School of Biomedical Engineering, ShanghaiTech University, Shanghai, China
wangqian2@shanghaitech.edu.cn
[2] School of Biomedical Engineering, Shanghai Jiao Tong University, Shanghai, China
[3] Department of Pathology, The Affiliated Drum Tower Hospital,
Nanjing University Medical School, Nanjing, China

Abstract. Deep learning has proven to be an effective approach to read whole-slide cytologic images for computer-assisted cervical cancer screening. To construct such a pipeline, it is often necessary to train a classifier, which decides if a patch (for example, sized 224 × 224) provides a positive cell reading or not. Following the clinical guidance, pathologists must label many such patches of both negative (N) and positive (P) cells. Then, a deep network can be trained for the binary N-P classification problem. In this paper, we take advantage of the complexity and illegibility of the cells that are negative according to current clinical guidance, to further improve the classification performance. We claim that the negative patches can be divided into two types: easy-negative (EN) and hard-negative (HN). The initial N-P binary classification can then be converted to an EN-HN-P triple-class problem. We also align the EN-HN and HN-P decision planes in parallel in the latent feature space where all input patches are encoded. The dual planes perform parallel classification then, following a well-planned curriculum learning scheme. Our results show that the proposed method can greatly enhance the performance of classifying positive patches and negative patches by using the better learned latent space and the related encoding of each patch.

Keywords: Cervical cancer · Image classification · Curriculum learning

1 Introduction

Cervical cancer is the second most common cancer among adult women in the world. A feasible way to screen cervical cancer is the thin-prep cytologic test (TCT) [7]. Specifically, the samples from the cervix are collected and sent to a lab, where pathologists examine the samples by microcopy or whole-slide imaging (WSI). They then make diagnoses according to the gold standard of the Bethesda

© The Author(s), under exclusive license to Springer Nature Switzerland AG 2022
S. Ali et al. (Eds.): CaPTion 2022, LNCS 13581, pp. 34–43, 2022.
https://doi.org/10.1007/978-3-031-17979-2_4

System (TBS) [9]. The screening technique is capable of finding early-stage abnormality of the positive cells, thus reducing the incidence and mortality rate of cervical cancer significantly [12]. However, considering the number of samples and the corresponding workload, it is challenging to widely adopt this screening system, especially in countries with limited healthcare resources.

Recently, deep learning has emerged as a promising tool for automatic cervical disease screening, which can effectively reduce the workload of pathologists. Concerning the complexity of the screening task, it is usually solved step by step. For example, Zhou et al. [17] proposed a three-step pipeline, which first starts to localize potentially abnormal cervical cells by object detection. Then, given the patch centered at the detected cell(s), a classification model is built to tell whether the patch is truly positive or not. Finally, the typical patches identified by the classification pass to the next stage, where the overall positive/negative diagnosis for the WSI of the subject is produced.

The above pipeline can efficiently handle the WSI data of extremely large size. Several existing works also adopt similar designs. Cao et al. [2] proposed the detection model namely AttFPN, which benefits from the clinical knowledge and the attention mechanism, to improve the detection performance. Cheng et al. [4] proposed a progressive method that combines multi-scale visual cues to identify abnormal cells. And a recurrent neural network (RNN) is adopted to complete the WSI-level classification based on the patches of the detected abnormal cells.

While the steps above are closely coupled, the patch-level classification has a significant role in the entire screening system. To screen cervical cancer, the detection model typically functions in a highly sensitive manner, such that the abnormal cells can be mostly detected even if they are rare in a large WSI. In this way, a slightly positive patient can still be diagnosed. However, the hypersensitive detection will produce many errors inevitably. And the patch-level classification, which further determines whether a detected cell (and its patch) is truly positive or negative, can refine the cell-level detection results.

The patch-level classification discussed in this paper falls into the category of image classification generally, as the patch is sized 224×224 (c.f. upper-left in Fig. 1). Such a patch is large enough to accommodate one to several cells, from which the positive or negative decision can be reached in reference to TBS. However, due to the complexity and illegibility of pathological images, there will inevitably be many hard images, and a few methods have been proposed to deal with this problem, such as [3,15,18]. In this paper, we propose that the negative samples can be further divided into easy negative (EN) and hard negative (HN), although they are grouped together in the clinically adopted TBS. Thus the original N-P binary classification can be converted to EN-HN-P classification. Moreover, we propose a parallel classification strategy, to generate dual classifiers in parallel for EN-HN and HN-P classification in the latent feature space. In this way, we can optimize the latent feature space and effectively improve the classification performance to separate the positive images from the negative.

The main contributions of our work are as follows.

- We split the N samples to EN and HN, and reshape a classical binary classification problem (N-P) into a triple-class problem (EN-HN-P).

- We further align the EN-HN and HN-P decision planes in parallel, and boost the intra-class compactness in the latent space, both of which contribute to the better encoding of the input images.
- We embed the above strategy into a curriculum learning [1] scheme, which adaptively enables the proposed constraints toward the high classification performance.

2   Method

2.1   Overview

We aim to solve a typical image classification problem in this paper. Specifically, we adopt the ResNet-style [6] backbone, which is a common choice for classifying medical images [14,16]. To take ResNet18 as the example of the encoder, the input images pass through the convolutional layers, pooling layers, and a global average pooling (GAP) layer, to arrive at the high-dimensional latent feature space (i.e., dim $= 512$ in our implementation). Next, we let the features go through two fully-connected (FC) layers to finally output the classification result.

Figure 1 shows the typical input images (upper-left in the figure), as well as illustrative distribution of the encoding in the latent space if following the conventional ResNet (upper-right, "Stage 1"). That is, the first FC layer appended to the encoder can be perceived as a linear decision plane applied to the latent feature space, which is denoted by the purple dashed line in the figure.

The modeling of the latent space is critical to deliver satisfactory classification performance. To this end, we treat the classical N-P binary classification described above as "Stage 1", and further propose "Stage 2" and "Stage 3" to train the classification network. Specifically, we separate the N-class inputs into EN and HN groups. After converting the N-P classification into EN-HN-P, we further embed the three stages into the unified curriculum learning scheme [10,14].

- **Stage 1: Binary N-P Classification.** We adopt the original N-P classification setting, e.g., to optimize the network with a binary cross-entropy loss.
- **Stage 2: Dual EN-HN and HN-P Classification.** Based on the network in Stage 1, we further split the FC layer, which is subsequent to the latent space, into two parallel ones. The optimization upon the two decision planes then helps reshape the latent feature space, as well as the encoding of the input images.
- **Stage 3: Intra-Class Compactness.** To further boost the encoding capability, we propose the intra-class compactness loss, such that the three groups of EN, HN, and P can distribute more compactly in the latent space. In this way, the decision planes between the separate groups can be more precisely estimated.

The details of the proposed method will be elaborated in the next.

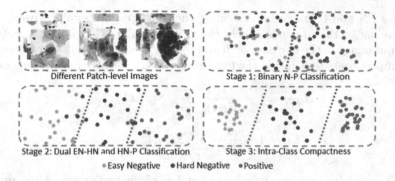

Different Patch-level Images   Stage 1: Binary N-P Classification

Stage 2: Dual EN-HN and HN-P Classification   Stage 3: Intra-Class Compactness

● Easy Negative ● Hard Negative ● Positive

Fig. 1. Our proposed method consists of three training stages. In Stage 1, we solve the problem as the classical N-P classification. In Stage 2, we split the N-P decision plane into two parallel ones for EN-HN and HN-P, respectively. In Stage 3, we further add the intra-class compactness constraint to the latent feature space. The typical input images for the three groups of EN, HN, and P are provided in the upper-left panel. The latent feature spaces, after training in three stages, are also illustrated in the figure. The dots of different colors represent images of different groups, and the dashed lines indicate the decision planes.

2.2   Dual Classifiers in Parallel

As the N-P binary classification becomes EN-HN-P classification, we split the original FC layer into two. The two new FC layers, corresponding to EN-HN (red dashed line in "Stage 2" of Fig. 1) and HN-P classification (purple dashed line) respectively, are parallel with each other. That is, we have

$$y_1 = w^\top x + b_1,$$
$$y_2 = w^\top x + b_2. \tag{1}$$

Here, w is the shared slope between the two FC layers, b_1, b_2 are respective biases, x is the image feature vector after GAP, and y_1, y_2 are produced by the two parallel FC layers. Since w is shared now, the two FC layers are parallel with each other in the latent space.

2.3   Intra-class Compactness

To distinguish the negatives and positives more clearly, we impose the constraint on the intra-class compactness to separate them further apart, thereby improving the classification performance. Specifically, we aim to make the intra-class distribution of EN and P more compact and the inter-class distance between them larger. The HN images then can fall in between the two groups then.

We first define the popular cosine similarity [11] between the feature vectors (x_i and x_j) of two input images:

$$S(x_i, x_j) = \frac{x_i^T \cdot x_j}{|x_i||x_j|} \tag{2}$$

Then, for EN and P images, we calculate their group centers respectively, using the encoding in the previous epoch. Next, we calculate the similarity between each input image in the current batch and the previously estimated group center. The compactness losses inside the two groups of EN and P are thus defined as

$$L_{EN} = 1 - \frac{1}{|\mathcal{B}_{EN}|} \sum_{i \in \mathcal{B}_{EN}} S(x_i, C_{EN}),$$

$$L_P = 1 - \frac{1}{|\mathcal{B}_P|} \sum_{i \in \mathcal{B}_P} S(x_i, C_P). \tag{3}$$

Here, $\mathcal{B}_{EN}$ and $\mathcal{B}_P$ indicate the current batches of the two groups, C_{EN} and C_P compute the center vectors for the two groups based on the previous epoch.

We further separate the EN and P classes by introducing $L_{EN-P} = S(C_{EN}, C_P)$. In this way, the two classes can move apart from each other in the latent space, while the HN group can distribute between them.

Finally, we derive the overall intra-class compactness loss:

$$L_{ICC} = L_{EN} + L_P + L_{EN-P}. \tag{4}$$

2.4   Implementation Details

As we choose ResNet as the backbone, our code follows the PyTorch-lightning framework [5]. In Stage 1, we use the binary cross-entropy loss for the binary N-P classification. In Stage 2, to deal with dual EN-HN and HN-P classification, we use two respective cross-entropy losses. In Stage 3, we add the intra-class compactness to the loss used in Stage 2.

We train all models for 150 epochs by using an Adam optimizer on one Nvidia 3060 GPU. And early stopping strategy is adopted, if the loss does not decrease in ten consecutive epochs. During training, we use the cosineannealing learning rate scheduler, with an initial learning rate of 1e-4 and batch size of 64. Besides, data augmentation is applied on the fly during model training, including random rotation, scaling, translation, and Gaussian noise, all models in our experiments is pretrained by ImageNet.

3   Experimental Results

3.1   Datasets

Patch-Level: We first prepare a set of 224×224 sized images to validate the proposed method. Specifically, we have applied an in-house cell detection tool to crop tens of thousands of images of both negative and positive [Anonymous Citation]. Then, a medical student has been instructed to label the cropped images, following the most strict criterion to identify the possibly positive. That is, she only labels those images as negative if she is fully confident; otherwise, the images are labeled as positive. Next, a senior pathologist (with 35-year experience) has examined these images, who is also aware of the student's labeling results. The senior pathologist sticks strictly to the TBS protocol and particularly tries the best to separate those images that are close to the N-P boundary. After the two-round labeling, we thus have three classes, i.e., EN for both negative labels, P for both positive, and HN for mismatched labeling. Note that the student and the senior pathologists do not know each other personally, and had no conflict of interest before this collaboration.

To ease the following comparison, we randomly draw from the above labeled database. In particular, we use 2517 EN, 2884 HN, and 2333 P images for this paper. The numbers of the three classes thus are largely balanced. A four-fold cross-validation will be conducted, and the data is split at the subject level in case a single subject may contribute multiple images or patches.

Subject-Level: In addition to patch-level validation, we further adopt two subject-level WSI datasets to demonstrate the merit of our method in real clinical use. The first dataset consists of 308 subjects (144 positives and 164 negatives), which means the two groups are largely balanced, we call it Dataset-1. The second dataset consists of 328 subjects (45 positives and 283 negatives), we call it Dataset-2. The positive samples account for 13.7% of the second dataset, which nearly reflects the real epidemiological incidence rate of cervical cancer [13].

For a certain subject, we detect 20 patches of the highest abnormal suspicion using the aforementioned cell detection tool. Then, those top-20 patches are classified by the proposed method. The final negative diagnosis is derived if all 20 classification tasks are negative; otherwise positive, the above ensembling strategy to the subject-level diagnosis is clinically feasible [17].

3.2   Classification Performance

Patch-Level: We perform patch-level cross-validation and verify the performance of individual classification methods with three different ResNet backbones (ResNet18, 34, and 50). The AUCs shown in Table 1 reflect the performance to separate the P images from EN+HN, which is the core of our task.

Table 1. Classification results (AUC) of comparing methods (B: binary classification; T: triple-class classification; P: parallel classification strategy; ICC: intra-class compactness; CL: three-stage curriculum learning strategy).

Method	ResNet18	ResNet34	ResNet50
B	0.8780 ± 0.0115	0.8805 ± 0.0115	0.8631 ± 0.0098
T	0.8794 ± 0.0112	0.8832 ± 0.0120	0.8862 ± 0.0096
B+P	0.8981 ± 0.0112	0.9001 ± 0.0118	0.8892 ± 0.0109
T+ICC	0.8698 ± 0.0119	0.8625 ± 0.0084	0.8773 ± 0.0119
B+P+ICC	0.9025 ± 0.0105	0.9032 ± 0.0110	0.8884 ± 0.0110
T+ICC+CL	0.8820 ± 0.0112	0.8837 ± 0.0115	0.8872 ± 0.0109
B+P+ICC+CL	$\mathbf{0.9112 \pm 0.0113}$	$\mathbf{0.9121 \pm 0.0115}$	$\mathbf{0.9023 \pm 0.0117}$

We treat the binary ("B") classification as the baseline. Meanwhile, as the N-P classification becomes EN-HN-P, one may naturally solve it in a three-class way ("T"). A direct comparison shows a similar performance of B and T, implying that simply converting a binary problem to triple cannot better classify a certain sub-group. On the contrary, by adopting the parallel classification (i.e., Stage 2, and "B+P" in Table 1), one may benefit from the more organized latent feature space and improve the classification performance.

The intra-class compactness further contributes to better classification performance, e.g., comparing "B+P" and "B+P+ICC". However, for "T" and "T+ICC", the AUCs drop for all three ResNet backbones. We argue that the intra-class compactness shall be carefully imposed. If the latent space has no prior knowledge of the EN-HN-P spectrum, the intra-class compactness loss will only be harmful.

Our proposed method integrates the dual classifiers and the intra-class compactness in the three-stage curriculum learning scheme, which corresponds to the last row ("B+P+ICC+CL") in Table 1. We note that the proposed method achieves the highest AUCs in all three backbones, compared to all other settings. It is also worth mentioning that the curriculum learning scheme benefits "T+ICC+CL" as well (i.e., versus "T+ICC"). The findings again underscore the importance of well-organized image encoding, as both "ICC" and "CL" contribute to shaping the latent space.

Subject-Level: The subject-level validation is performed on two datasets, while the results are provided in Table 2. Note that Dataset-1 is largely balanced in the positive/negative subjects; thus we focus on its ACC for the subject-level classification result. It is observable that our method yields a much higher ACC than the binary classification via ResNet34. Here, for fair comparison, our method also adopts the same ResNet34 backbone.

In Dataset-2, we focus on precision and recall. The reason is that Dataset-2 reflects the real epidemiological incidence rate. From the table, we observe the proposed method keeps a slightly higher precision than the baseline. Meanwhile, its recall has increased significantly. That is, the proposed patch-level classification method leads to equally or more sensitive screening of cervical cancer at the subject level, whereas it can reduce many false positives and potentially ease the workload of the pathologists.

Table 2. Classification results of the subject-level validation.

Dataset-1	AUC	ACC	Precision	Recall	F1-score
Binary (ResNet34)	0.7916	0.5779	0.8036	0.2744	0.4091
Proposed Method	**0.8297**	**0.6948**	**0.8302**	**0.5366**	**0.6519**
Dataset-2	AUC	ACC	Precision	Recall	F1-score
Binary (ResNet34)	0.7702	0.3140	0.9531	0.2155	0.3516
Proposed Method	**0.8065**	**0.5183**	**0.9630**	**0.4594**	**0.6220**

3.3   Evolving of the Latent Space

To verify the image encoding, we use t-SNE [8] to visualize the latent space after individual training stages. As shown in Fig. 2, when only a binary classifier is used for the N-P classification (Stage 1), we can find that the images are highly mixed in the latent space. There is no clear boundary between the positive (red dots) and the negative (green and blue combined). It is also worth noting that the blue dots (for HN) are generally closer to the red (P), in comparison to green (EN). It confirms that the EN-HN-P distribution can be a spectrum-like continuum, thus we can take advantage of the distribution for better classification. Next, in Stage 2, the two parallel classifiers place additional constraint to the encoding process, which effectively separates the three groups of EN, HN, and P. In particular, the boundary between HN and P becomes much clearer, implying higher accuracy in identifying positive images from the entire dataset.

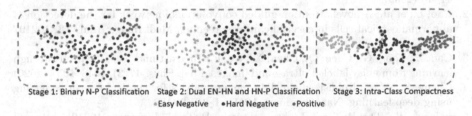

Stage 1: Binary N-P Classification   Stage 2: Dual EN-HN and HN-P Classification   Stage 3: Intra-Class Compactness
• Easy Negative   • Hard Negative   • Positive

Fig. 2. Visualization of the latent feature space. The green, blue, and red dots represent the EN, HN, and P images, respectively. (Color figure online)

When the intra-class compactness is further introduced in Stage 3, we observe that the dots of different colors are tightly distributed. Note that, although we have no direct restriction to the compactness of HN, still the blue dots are mapped in-between the EN and P images following our expectation. In general, the classification task can be more easily solved by the image encoding in Stage 3.

4   Discussion and Conclusion

We proposed a parallel classification framework, which carefully encodes the input images and generates two classifiers for EN-HN and HN-P decision planes. By separating the N images into EN and HN, we demonstrate that the dual classifiers, intra-class compactness constraint, and curriculum learning strategy in-together can yield higher accuracy in separating the P images from the rest.

Our method has a high potential for cervical cancer screening. Given the good performance on patch-level classification, one may integrate our method with a sophisticated subject-level ensembling framework. In this way, the entire screening system can deliver better performance to meet the clinical demands. Also, our method is not restricted to TCT images only; it can be extended to many more applications of pathological images.

Our method has some limitations, which will be addressed in future work. While we recruit multiple pathologists to label the data, it is possible to use a roughly trained model to identify those hard images by itself. Then, the model can evolve, e.g., following an active learning scheme. Moreover, being specific to cervical cancer screening, we will utilize the ambiguity of the cells that lie on the positive side. That is, the hard-positive (HN) images, as well as the detailed positive grading in TBS, can all help train the network toward precise diagnosis instead of just screening, and we can also recruit more pathologists to have more labels.

References

1. Bengio, Y., Louradour, J., Collobert, R., Weston, J.: Curriculum learning. In: Proceedings of the 26th Annual International Conference on Machine Learning, pp. 41–48 (2009)
2. Cao, L., et al.: A novel attention-guided convolutional network for the detection of abnormal cervical cells in cervical cancer screening. Med. Image Anal. **73**, 102197 (2021)
3. Cao, Z., Yang, G., Chen, Q., Chen, X., Lv, F.: Breast tumor classification through learning from noisy labeled ultrasound images. Med. Phys. **47**(3), 1048–1057 (2020)
4. Cheng, S., et al.: Robust whole slide image analysis for cervical cancer screening using deep learning. Nat. Commun. **12**(1), 1–10 (2021)
5. Falcon, W.: The PyTorch Lightning team: PyTorch Lightning (2019). https://doi.org/10.5281/zenodo.3828935. https://github.com/PyTorchLightning/pytorch-lightning
6. He, K., Zhang, X., Ren, S., Sun, J.: Deep residual learning for image recognition. In: Proceedings of the IEEE Conference on Computer Vision and Pattern Recognition, pp. 770–778 (2016)
7. Koss, L.G.: The papanicolaou test for cervical cancer detection: a triumph and a tragedy. JAMA **261**(5), 737–743 (1989)
8. Van der Maaten, L., Hinton, G.: Visualizing data using t-SNE. J. Mach. Learn. Res. **9**(11) (2008)
9. Nayar, R., Wilbur, D.C.: The Bethesda System for Reporting Cervical Cytology: Definitions, Criteria, and Explanatory Notes. Springer, Cham (2015). https://doi.org/10.1007/978-3-319-11074-5
10. Nebbia, G., et al.: Radiomics-informed deep curriculum learning for breast cancer diagnosis. In: de Bruijne, M., et al. (eds.) MICCAI 2021. LNCS, vol. 12905, pp. 634–643. Springer, Cham (2021). https://doi.org/10.1007/978-3-030-87240-3_61
11. Nguyen, H.V., Bai, L.: Cosine similarity metric learning for face verification. In: Kimmel, R., Klette, R., Sugimoto, A. (eds.) ACCV 2010. LNCS, vol. 6493, pp. 709–720. Springer, Heidelberg (2011). https://doi.org/10.1007/978-3-642-19309-5_55
12. Schiffman, M., Castle, P.E., Jeronimo, J., Rodriguez, A.C., Wacholder, S.: Human papillomavirus and cervical cancer. The Lancet **370**(9590), 890–907 (2007)
13. Vaccarella, S., Lortet-Tieulent, J., Plummer, M., Franceschi, S., Bray, F.: Worldwide trends in cervical cancer incidence: impact of screening against changes in disease risk factors. Eur. J. Cancer **49**(15), 3262–3273 (2013)

14. Wei, J., et al.: Learn like a pathologist: curriculum learning by annotator agreement for histopathology image classification. In: Proceedings of the IEEE/CVF Winter Conference on Applications of Computer Vision, pp. 2473–2483 (2021)

15. Xue, C., Dou, Q., Shi, X., Chen, H., Heng, P.A.: Robust learning at noisy labeled medical images: applied to skin lesion classification. In: 2019 IEEE 16th International Symposium on Biomedical Imaging (ISBI 2019), pp. 1280–1283. IEEE (2019)

16. Zhang, Q., et al.: A GPU-based residual network for medical image classification in smart medicine. Inf. Sci. **536**, 91–100 (2020)

17. Zhou, M., et al.: Hierarchical pathology screening for cervical abnormality. Comput. Med. Imaging Graph. **89**, 101892 (2021)

18. Zhu, C., Chen, W., Peng, T., Wang, Y., Jin, M.: Hard sample aware noise robust learning for histopathology image classification. IEEE Trans. Med. Imaging **41**(4), 881–894 (2021)

Detection and Diagnosis

Lightweight Transformer Backbone
for Medical Object Detection

Yifan Zhang[1,2]($\boxtimes$), Haoyu Dong[1], Nicholas Konz[3], Hanxue Gu[3],
and Maciej A. Mazurowski[1,3,4,5]

[1] Department of Radiology, Duke University, Durham, USA
yifan.zhang.2@vanderbilt.edu, {hd108,maciej.mazurowski}@duke.edu
[2] Department of Computer Science, Vanderbilt University, Nashville, USA
[3] Department of Electrical and Computer Engineering, Duke University,
Durham, USA
{nicholas.konz,hg119}@duke.edu
[4] Department of Biostatistics and Bioinformatics, Duke University, Durham, USA
[5] Department of Computer Science, Duke University, Durham, USA

Abstract. Lesion detection in digital breast tomosynthesis (DBT) is an important and a challenging problem characterized by a low prevalence of images containing tumors. Due to the label scarcity problem, large deep learning models and computationally intensive algorithms are likely to fail when applied to this task. In this paper, we present a practical yet lightweight backbone to improve the accuracy of tumor detection. Specifically, we propose a novel modification of visual transformer (ViT) on image feature patches to connect the feature patches of a tumor with healthy backgrounds of breast images and form a more robust backbone for tumor detection. To the best of our knowledge, our model is the first work of Transformer backbone object detection for medical imaging. Our experiments show that this model can considerably improve the accuracy of lesion detection and reduce the amount of labeled data required in typical ViT. We further show that with additional augmented tumor data, our model significantly outperforms the Faster R-CNN model and state-of-the-art SWIN transformer model.

1   Introduction

Medical Object Detection (OD) distinguish the object of interest from medical images, which is important in the downstream medical applications such as diagnosis. In the era of big data, hospitals are able to gather a large amount of images to train detection models and use them to assist radiologist with different disciplines [17]. However, training such models usually needs accurate tumor bounding boxes, which is labor-intensive and resource consuming to annotate.

Y. Zhang and H. Dong—Equal contribution.

Supplementary Information The online version contains supplementary material available at https://doi.org/10.1007/978-3-031-17979-2_5.

© The Author(s), under exclusive license to Springer Nature Switzerland AG 2022
S. Ali et al. (Eds.): CaPTion 2022, LNCS 13581, pp. 47–56, 2022.
https://doi.org/10.1007/978-3-031-17979-2_5

The scarcity of available bounding boxes constrains the volume of models and limits the overall model performance, eliciting the need for an innovative structure to perform effective medical object detection with the most efficient model design.

Having a reliable model architecture for medical OD is essential. The classic methods for object detection utilize Convolutional Neural Networks (CNN) [1, 3,4] to select a considerable number of regions for location prediction. To reduce the number of assigned areas, Region-Based Convolutional Neural Networks (R-CNN) [8] choose an exact number of proposed regions into CNN to improve the model efficiency. To further solve the drawbacks of computational efficiency in R-CNN, Fast R-CNN [7] and Faster R-CNN [14] were proposed to feed the input image directly into CNN to generate a convolutional feature map for proposal regions. Looking at the complete picture and predicting a class probability in each grid, YOLO [13] outperforms the previous methods and becomes the most efficient and effective framework in CNN-based detection.

Recent studies in Transformer [15], an attention based neural network structure, have advanced the performance in OD. With the evolution of Transformer on vision, Visual Transformer (ViT) [6] is proposed to model long-term dependencies of image patches. The latest state-of-the-art (SOTA) Transformer method on vision, called SWIN Transformer [12], uses shifted windows to construct hierarchical visual representations for downstream application and is beneficial for most modeling in natural images. However, ViT and SWIN are constrained by a considerable amount of required data, which is usually unreachable in medical object detection scenarios. For example, both methods targeted on COCO [11], which consists of 118K labelled training samples, while most medical dataset consists of less than 1K labelled samples.

To improve the performance of CNN-based medical OD models and address the data hunger in the attention mechanism, we introduce a lightweight Transformer backbone for improving detection accuracy without extra annotations. Specifically, we replace the feature pyramid network (FPN) module by using a weighted sum strategy to integrate the features from different layers instead of summing them equally. To achieve this goal, we use feature rearrangement and reconstruction to reshape and restore the feature maps of ResNet into and from feature patches, which unifies the model on outputs of each ResNet layer with only one ViT layer. The reconstruction task also allows the model to fully utilize the training data. Besides, we introduce novel lightweight attention of ViT to enhance the representations of rearranged feature patches. The experiments demonstrate that using our model for medical object detection is highly promising, achieving significantly better performance to the Faster R-CNN method. It also considerably outperforms the SOTA ViT model (SWIN).

2   Methodology

In this paper, we intend to provide a lightweight Transformer backbone for medical object detection with limited positively labelled data. This section presents the key components of the proposed method.

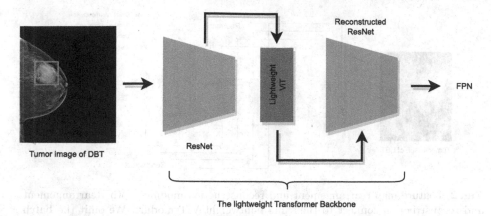

Fig. 1. Illustration of our lightweight Transformer backbone. The ResNet and reconstructed ResNet have the same shape. The black arrows represent information flow.

2.1 Overview of Proposed Method

As shown in Fig. 1, we propose a new lightweight ViT backbone for coupling with the feature pyramid network in order to improve detection performance. In our pipeline, raw images are fed into a ResNet [9] to generate feature maps corresponding to the activation map of each hidden layer. Because spatial attention can significantly improve the connections between pixels, before feeding these out-puts into the next FPN [10] that uses multi-scale pyramidal hierarchy to construct feature pyramids for Region Proposal Network (RPN) and Region of Interest (RoI) pooling [14], we apply attention on image feature patches of the outputs of ResNet to improve the hidden representations of each inputs to FPN.

2.2 Feature Map Rearrangement and Reconstruction

In this section, we introduce how the feature maps of ResNet are rearranged as feature patches and fit the inputs of our ViT, and how the feature patches are reconstructed with the original shape of feature maps generated by ResNet. The feature rearrangement and reconstruction process of our model are shown in Fig. 2.

Feature Map Rearrangement. The ResNet outputs have the shapes of (B, C_k, H_k, W_k), in which B is the batch size, C_k, H_k and W_k denote the number of channels, feature map height and width in the kth layer. Each patch of feature map in the ResNet outputs will be subjected to the following rearrangement in the first section:

$$z_k = [x_{kp}^1 E; x_{kp}^2 E; ...; x_{kp}^{N_k} E],\qquad(1)$$

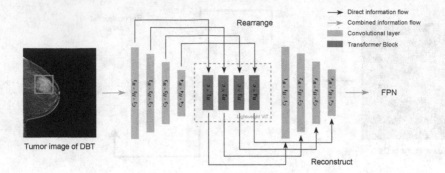

Fig. 2. Feature map rearrangement and reconstruction module. Both Rearrangement and reconstruction connect to the same lightweight ViT module. We omit the batch size for clarity.

where x_{kp} and z_k stand for patches of feature map x and feature patches z in the kth layer. E denotes a *feature map rearrangement embedding*, $E \in \mathbb{R}^{(w \cdot h \cdot C_k) \times c}$, in which w and h are width and height of a single patch.

The feature map transformation embedding is for transforming an original feature map to a feature map on patches. To be more specific, it rearranges the shape of feature map by the formula below:

$$(B, C_k, H_k, W_k) \rightarrow (B, n_k, c). \tag{2}$$

Here n_k is the number of feature patches in k^{th} layer, and C_k is the number of channels in the current hidden layer, which is computed as

$$C_k = c \cdot 2^{(k-1)}, c = 256. \tag{3}$$

Since the output of ResNet follows a pyramid structure, with the increment in hidden dimension and stride over spatial dimension, we use divisible numbers of the spatial size of the last-layer representation as the patch size, which is (5, 4). Then, all shallow layers are first resized along hidden dimension to match the depth information, i.e., $(256 \times 2, n) \rightarrow (256, (2 \times n))$. Therefore, the number of feature patches for the kth layer will be

$$n_k = (H_k/h) \cdot (W_k/w) \cdot 2^{(k-1)}, h = 5, w = 4. \tag{4}$$

Feature Map Reconstruction. After the rearrangement and the lightweight Transformer module, feature patches will be reconstructed to their original shape as shown in the formula

$$(B, n_k, c) \rightarrow (B, C_k, H_k, W_k). \tag{5}$$

Because the only quantity that changes across layers is the total number of feature patches, this design ensures weight sharing of the ViT module.

2.3   Lightweight Transformer on Feature Patches

We design a lightweight Transformer module to enhance the representation of feature patches, consisting of positional embedding, attention, and feedforward components. The lightweight ViT module in our model is illustrated in Fig. 3.

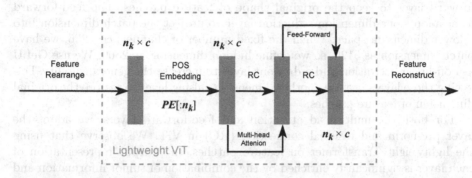

Fig. 3. The lightweight Transformer module. It consists of positional embedding, attention and feed-forward. All vectors omit the batch size dimension for clarity.

Positional Embedding. After the feature rearrangement, feature patches has the shape of (B, n_k, c). As with BERT [5] and ViT [6], we append a learnable positional embedding E_{pos} to assist the network in remembering the locations of individual patches, as

$$z_k = z_k + E_{pos}[: n_k], \tag{6}$$

where E_{pos} is a shared positional embedding for feature patches of each rearranged ResNet feature map. It has a maximum length of maximal number of feature patches $N \cdot 2^{min(k)-1}$, where N denotes the number of patches in the first ResNet layer, which is 4096, and $min(k) = 1$.

Multi-head Attention. We adopt the multi-head self-attention mechanism in ViT to jointly infer attention from different representation subspaces. The output of the self-attention is a scaled dot-product:

$$Attention(Q, K, V) = softmax(\frac{QK^T}{\sqrt{d_k}})V, \tag{7}$$

where $Q, K, V \in \mathbb{R}^{(w \cdot h \cdot c)}$ are query, key and value embeddings, and $\sqrt{d_k}$ is the dimension of the key vector k and query vector q. We extend it to the multi-head attention:

$$MultiHead(Q, K, V) = Concat(head_1, ..., head_I)W^O, \tag{8}$$

where

$$head_i = Attention(QW_i^Q, KW_i^K, VW_i^V), \tag{9}$$

Here W_i^Q, W_i^K, W_i^V, W^O denote trainable parameters corresponding to Q, K, V in the i_{th} attention head, and the output. In this study, we use $I = 8$. The results from multiple heads are concatenated and then transformed with a feed-forward network.

Feed-Forward. We use the same input and output dimension in the feed-forward layer to keep the original shape of feature patches. The feed-forward layer adopts one dimension reduction layer to project the patch dimension into a lower dimension space. When the fixed number of channels $c = 256$, we have patch dimension as 5120, so we define hidden dimension as 2560. We use GeLU as non-linear activation and a dropout layer to increase the generalizability. Following those layers, we use another dimension-raising layer to restore the original dimension of feature patches.

For both the multi-head attention and feed-forward layers, we adopt the layer pre-norm and residual connection (RC) in ViT. We observe that using the lightweight Transformer on feature patches, the output representation of each layer is significantly enriched by the combination of tumor information and different locations in medical images.

3 Experiments and Results

3.1 Dataset and Evaluation Metrics

We use a publicly available dataset of breast cancer screening scans: the Digital Breast Tomosynthesis (BCS-DBT) dataset [2]. The BCS-DBT dataset comprises cancer cases that are normal, actionable, non biopsy-proven, and biopsy-proven. It contains 22032 breast tomosynthesis scans from 5060 individuals, with each scan containing up to 4 anatomical views and dozens of spatially-aligned slices in each view. Figure 4 shows three examples of DBT images.

In our study, we use only the tumor slices with bounding boxes. In BCS-DBT, there are 299 tumor slices with 346 bounding boxes. As indicated in [2], we split the data into a training set with 233 tumor slices and 274 bounding boxes, and a validation set with 75 tumor slices and 75 bounding boxes.

We use the AP (Average Precision) metrics for object detection as a quantitative study of our model. There are metrics relying on IoU (Intersection over Union) that describes the intersection of ground-truth bounding boxes and predicted bounding boxes from models. The IoU formula is known as

$$IoU(A, B) = \frac{(A \cap B)}{(A \cup B)}, \tag{10}$$

where A and B are ground-truth bounding boxes and preidcted bounding boxes from models, respectively. $IoU(A, B) \in [0, 1]$.

In our study, we employ AP, AP50, AP75, APm, and APl as analysis criteria, with AP50 serving as an indicator of model performance. Here, AP50 and AP75 counts the samples that have at least 0.5 and 0.75 IoU areas respectively and AP

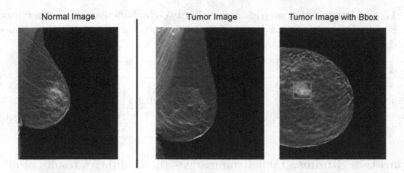

Fig. 4. The examples of normal and tumor images from BCS-DBT. Res box denotes the bounding box of the tumor image. (Color figure online)

is the average of AP50 to AP90 with step size 5. APm and APl are for medium objects with areas $\in [32^2, 96^2]$ and large objects with areas more than 96^2.

3.2 Implementation Details

We use Detectron2 [16] as our object detection framework. In our study, we implement two comparative models to demonstrate the competitiveness and generalization ability of our model. We use the other official implementations and default hyperparameters for training all three models in the Detectron2.

Faster R-CNN. We adopt the Faster R-CNN in Detectron2 using CNN as the backbone feature extractor. It utilizes a pretrained ResNet50 to get feature maps. During training, we set the batch size of 4 for the best performance.

SWIN Transformer. We implement a pretrained SWIN Transformer as the feature extractor of a Faster R-CNN model. It utilizes a windows size of 7 and an embedding dimension of 96. During training, we set the batch size of 4 and the learning rate of 0.001 for the best performance.

The original Detectron2 framework has random data augmentation that resizes images to 8 various shapes, some of which may be incompatible with our preset patch size. The widths and heights of hidden feature maps should be an integer multiple of the width and heights of the preset patch. As a result, we remove the random data resize in the experiments.

3.3 Experimental Results

We compare our proposed lightweight Transformer backbone to Faster R-CNN and SWIN Transformer in tumor detection and evaluate the performance of the models. We further prove the effectiveness of our model with additional augmented tumor data. The detailed method for tumor augmentation is attached in the appendix for clarity.

Table 1. The quantitative metrics of our lightweight backbone and other methods (mean ± std). Each simulation was performed 5 times for computing the means and standard deviation of criteria.

Method	AP	AP50	AP75	APm	APl
Faster R-CNN	11.78 (±1.08)	39.55 (±1.15)	4.06 (±2.24)	**9.36** (±3.12)	12.12 (±1.03)
SWIN transformer	4.28 (±0.85)	14.29 (±1.38)	2.01 (±0.97)	1.44 (±0.96)	4.71 (±0.75)
Lightweight (Ours)	**13.71** (±1.20)	**42.04** (±2.74)	**4.73** (±1.66)	6.20 (±3.07)	**14.45** (±1.62)

Comparative Studies. Table 1 summarizes the qualitative results of all four methods in tumor detection performance. The AP50 shows that our lightweight Transformer approach has achieved significantly more accurate detection of tumors, improving 7.2% (+2.49) on Faster R-CNN. The standard deviations show that our Lightweight Transformer method has more stability in strict criterion AP75 and medium object detection APm, whereas relatively diverse in other criteria, including AP, AP50, and APl.

On the contrary, the SWIN Transformer backbone performs considerably worse than both Faster R-CNN and Lightweight Transformer backbones. The AP50 shows that the performance of SWIN Transformer backbone drops by 25.26 and 27.75 comparing with Faster R-CNN and Lightweight Transformer backbones respectively, indicating the ineffectiveness of direct application of Transformer image feature extraction due to the scarity of available data.

Table 2. The main metrics and their difference of our lightweight backbone and other methods on augmented tumor dataset (mean ± std). Each simulation was performed 5 times for computing the means and standard deviation of criteria.

Method	AP	AP50	AP Change	AP50 Change
Faster R-CNN	12.84 (±2.36)	41.42 (±1.96)	+1.06	+1.87
SWIN transformer	4.18 (±1.08)	14.22 (±1.90)	−0.1	−0.07
Lightweight (Ours)	**13.41** (±0.82)	**44.03** (±2.12)	−0.3	+1.99

Evaluation on Augmented Dataset. We further compare our method with Faster R-CNN and SWIN using a 4x larger augmented training dataset through inserting tumors into normal images. The details of insertion can be found in the appendix. This leads to 932 tumor slices and 973 bounding boxes for training, while the validation set is kept the same. Table 2 shows that with additional labeled data, the lightweight Transformer method performs better than Faster R-CNN baseline in both AP (13.41) and AP50 (44.03), with higher increment (+1.99) in AP50 as well.

At the same time, the performance of the SWIN Transformer doesn't change much on AP (−0.1) and AP50 (−0.07), showing that the direct application of Transformer-based models as feature extractors needs a considerably more extensive dataset.

4   Conclusion

We proposed the lightweight Transformer backbone in this work to improve the performance of the medical object detection model in the context of breast tumor detection. As a novel backbone for improving high-resolution breast tumor detection's performance and stability and achieving higher performance on existing backbones without extra tumor annotations, our techniques provides a new idea for applying attention to related problems. We further prove that the direct application of Transformer-based methods on medical object detection requires a larger dataset, demonstrating the advantages of our proposed method.

References

1. Albawi, S., Mohammed, T.A., Al-Zawi, S.: Understanding of a convolutional neural network. In: 2017 International Conference on Engineering and Technology (ICET), pp. 1–6. IEEE (2017)
2. Buda, M., et al.: Data from the breast cancer screening-digital breast tomosynthesis (BCS-DBT). Data from The Cancer Imaging Archive (2020)
3. Cai, Z., Vasconcelos, N.: Cascade R-CNN: delving into high quality object detection. In: Proceedings of the IEEE Conference on Computer Vision and Pattern Recognition, pp. 6154–6162 (2018)
4. Chen, Y., Li, W., Sakaridis, C., Dai, D., Van Gool, L.: Domain adaptive faster R-CNN for object detection in the wild. In: Proceedings of the IEEE Conference on Computer Vision and Pattern Recognition, pp. 3339–3348 (2018)
5. Devlin, J., Chang, M.W., Lee, K., Toutanova, K.: Bert: pre-training of deep bidirectional transformers for language understanding. arXiv preprint arXiv:1810.04805 (2018)
6. Dosovitskiy, A., et al.: An image is worth 16x16 words: transformers for image recognition at scale. arXiv preprint arXiv:2010.11929 (2020)
7. Girshick, R.: Fast R-CNN. In: Proceedings of the IEEE International Conference on Computer Vision, pp. 1440–1448 (2015)
8. Girshick, R., Donahue, J., Darrell, T., Malik, J.: Rich feature hierarchies for accurate object detection and semantic segmentation. In: Proceedings of the IEEE Conference on Computer Vision and Pattern Recognition, pp. 580–587 (2014)
9. He, K., Zhang, X., Ren, S., Sun, J.: Deep residual learning for image recognition. In: Proceedings of the IEEE Conference on Computer Vision and Pattern Recognition, pp. 770–778 (2016)
10. Lin, T.Y., Dollár, P., Girshick, R., He, K., Hariharan, B., Belongie, S.: Feature pyramid networks for object detection. In: Proceedings of the IEEE Conference on Computer Vision and Pattern Recognition, pp. 2117–2125 (2017)
11. Lin, T.-Y., et al.: Microsoft COCO: common objects in context. In: Fleet, D., Pajdla, T., Schiele, B., Tuytelaars, T. (eds.) ECCV 2014. LNCS, vol. 8693, pp. 740–755. Springer, Cham (2014). https://doi.org/10.1007/978-3-319-10602-1_48
12. Liu, Z., et al.: Swin transformer: hierarchical vision transformer using shifted windows. In: Proceedings of the IEEE/CVF International Conference on Computer Vision, pp. 10012–10022 (2021)
13. Redmon, J., Divvala, S., Girshick, R., Farhadi, A.: You only look once: unified, real-time object detection. In: Proceedings of the IEEE Conference on Computer Vision and Pattern Recognition, pp. 779–788 (2016)

14. Ren, S., He, K., Girshick, R., Sun, J.: Faster R-CNN: towards real-time object detection with region proposal networks. In: Advances in Neural Information Processing Systems, vol. 28 (2015)
15. Vaswani, A., et al.: Attention is all you need. In: Advances in Neural Information Processing Systems, vol. 30 (2017)
16. Wu, Y., Kirillov, A., Massa, F., Lo, W.Y., Girshick, R.: Detectron2 (2019). https:// github.com/facebookresearch/detectron2
17. Yang, R., Yu, Y.: Artificial convolutional neural network in object detection and semantic segmentation for medical imaging analysis. Front. Oncol. 11, 573 (2021)

Contrastive and Attention-Based Multiple Instance Learning for the Prediction of Sentinel Lymph Node Status from Histopathologies of Primary Melanoma Tumours

Carlos Hernandez Perez[1]([📧]) [ID], Marc Combalia Escudero[2,3] [ID],
Susana Puig[2,3] [ID], Josep Malvehy[2,3] [ID], and Veronica Vilaplana Besler[1] [ID]

[1] Image Processing Group, Universitat Politècnica de Catalunya, Barcelona, Spain
{carlos.hernandez.p,veronica.vilaplana}@upc.edu
[2] Dermatology Department, Melanoma Unit, Hospital Clínic de Barcelona,
IDIBAPS, University of Barcelona, Barcelona, Spain
[3] Athena Tech, S.L., Barcelona, Spain

Abstract. Sentinel lymph node status is a crucial prognosis factor for melanomas; nonetheless, the invasive surgery required to obtain it always puts the patient at risk. In this study, we develop a Deep Learning-based approach to predict lymph node metastasis from Whole Slide Images of primary tumours. Albeit very informative, these images come with complexities that hamper their use in machine learning applications, namely their large size and limited datasets. We propose a pre-training strategy based on self-supervised contrastive learning to extract better image feature representations and an attention-based Multiple Instance Learning approach to enhance the model's performance. With this work, we quantitatively demonstrate that combining both methods improves various classification metrics and qualitatively show that contrastive learning encourages the network to output higher attention scores to tumour tissue and lower scores to image artifacts.

Keywords: Whole slide image · Contrastive learning · Attention-based Multiple Instance Learning · Early detection

1 Introduction

Skin cancer incidence has increased in the last decade worldwide [1], and it's the fourth leading cause of cancer-related mortality [16]. Melanoma constitutes the leading cause of death due to skin cancer when considering its incidence

Work supported by the Spanish Research Agency (AEI) under project PID2020-116907RB-I00 of the call MCIN/AEI/10.13039/501100011033 and the project 718/C/2019 funded by Fundació la Marato de TV3.

© The Author(s), under exclusive license to Springer Nature Switzerland AG 2022
S. Ali et al. (Eds.): CaPTion 2022, LNCS 13581, pp. 57–66, 2022.
https://doi.org/10.1007/978-3-031-17979-2_6

and mortality. Therefore, there is a need for early detection of melanoma: the five-year survival rate of melanoma rapidly decreases as its stage advances, from 97% for stage I to only 15–20% for stage IV.

The result of a sentinel lymph node biopsy (SLNB) is a key early prognosis factor for melanoma patients. An SLNB is a procedure in which the sentinel lymph node is identified, removed, and examined to determine whether cancer cells are present. Only patients that have already been diagnosed with cancer undergo this surgery. A negative SLNB result suggests that cancer has not yet spread to nearby lymph nodes or other organs. However, 80% of patients do not benefit from this intervention because they have unaffected lymph nodes [7]. Thus, it would be desirable to determine the likelihood of a positive SLNB in advance to reduce the number of unnecessary surgeries and their associated morbidities.

Artificial intelligence and, especially, Convolutional Neural Networks (CNN) [8] have proved to be state-of-the-art methods for medical imagining analysis [15,18–20]. In recent years, the value of deep learning empowered computer-assisted diagnosis has been established in dermatological imaging-based decision-making models [5]. It has been successfully implemented in a variety of use cases, from finding skin lesion biomarkers [6] to identifying tumour tissue in Whole Slide Images (WSI) [13]. Unfortunately, the literature on the prediction of SLNB positivity from WSI of primary cutaneous melanoma tumours is scarce. To our knowledge, the only work that addresses this problem is Brinker et al. [3] where they use a Multiple Instance Learning (MIL) framework combining clinical data, cell features, and the slide feature vectors with a squeeze and excitation block before classification. Our approach tackles the same problem by combining self-supervised contrastive learning [11] to improve the quality of the extracted features and an attention-based MIL strategy to improve the model's classification capabilities. Additionally, the attention mechanism allows the visualization of the relative patch importance for the final model's prediction.

2   Materials and Methods

In this section, we first present the dataset and briefly discuss the usage of MIL in our work. Afterwards, we describe the details of our proposed model, and, finally, we present a primer on contrastive learning and how we apply it to WSI analysis.

2.1   Dataset

Ethics approval was obtained from the ethics committee of Hospital Clinic de Barcelona before the study was initiated. A total of 195 digitised WSIs were used, each coming from a different patient. WSIs were obtained from a cutaneous biopsy of a primary melanoma tumour, while the target variable was obtained after analyzing an SLNB (107 SN negative and 88 SN positive). Both biopsies were performed, at most, within a year. Available clinical data included the patient's age, tumour thickness, and ulceration, presented in Table 1.

Table 1. Clinical data. Mean and standard deviation computed for the age and tumour thickness.

Clinical variable	SLN+	SLN−	All
Age (years)	54.2 ± 14.6	58.1 ± 14.6	56.3 ± 14.7
Ulceration (yes/no)	43/45	64/43	107/88
Tumour thickness (mm)	1.12 ± 0.69	1.26 ± 0.65	1.2 ± 0.68
SLN status	88	107	195

Patch Extraction: The tissue was segmented from the background through Otsu's thresholding method [14]. Afterwards, patches of 256×256 pixels were cropped without any overlap. Then, patches containing a small tissue area were regarded as non-informative and discarded from the dataset. The average number of extracted patches per slide was 488.

2.2 Multiple Instance Learning

Since only labels at the WSI level are available (the positivity of the SLNB), we used a weakly supervised architecture based on Multiple Instance Learning. MIL is a family of weakly-supervised learning algorithms in which the learner receives a set of labeled bags where each contains an unconstrained number of unlabeled instances. In our case, bags are slides, S, and instances are patches, $\mathcal{X}$. An assumption needs to be made regarding the relationship between a bag, its instances, and the class label of the bag. Two main assumptions are tested in this work:

The Standard Assumption: Each patch $x \in \mathcal{X}$ has an associated label $y \in \{0,1\}$ which is hidden to the learner. If at least one of them has a positive label, the whole slide is considered to be positive. Formally, let $S = \{(x_1, y_1), \ldots, (x_K, y_K)\}$ be a slide represented as a set of patches with their labels. The label of S is then $c(S) = 1 - \prod_{i=1}^{K}(1 - y_k)$.

Attention-Based MIL Pooling: In contrast to the previous assumption, attention mechanisms allow for a more flexible and adaptive MIL pooling [10] through their ability to adjust to a task and data by having trainable parameters. Let $H = \{h_1, ..., h_K\}$ be a slide represented with a set of K patch feature embeddings, being $h_i = e(x_i)$ where e is any encoder; then a MIL pooling can be expressed as:

$$\mathbf{r} = \sum_{k=1}^{K} a_k \mathbf{h_k}; \quad \mathbf{a} = \{a_1, ..., a_K\}; \quad \sum_{k=1}^{K} a_k = 1 \qquad (1)$$

where a_k is the attention weight for the k-th patch and $\mathbf{r}$ is the aggregated patch features for the slide. Let The label of S is $c(S) = m(\mathbf{r})$, where $m(\cdot)$ is typically a multilayer perceptron (MLP).

2.3   Proposed Model

Our proposed method expands the work by Ming Y. Lu et al. [13] and consists of two steps: a feature extractor and a classification module, as shown in Fig. 1. The feature extractor uses the convolutional layers of a ResNet50 model to extract features from the patches, reducing the dimensionality of the data. This, in turn, enables the classification module to fit all the slide's info into a single commercial GPU. During classification, for both training and inference, the model examines and ranks all patches in the tissue regions of a WSI, assigning an attention score to each patch, which informs its contribution or importance to the collective slide-level representation for each class. Attention scores are obtained with a gated attention mechanism [10]. Next, each feature vector is forwarded through its respective class branch consisting of an MLP with a single neuron as an output. Finally, both branch outputs are forwarded together through a softmax layer. The maximum of the two values is taken as the prediction for the WSI.

Clinical data is added through an additional input head with two linear layers and non-linearities. The resulting vector and the aggregated patch features are normalized before concatenation. The combined feature vector is then forwarded through each class branch.

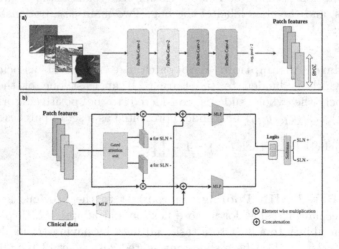

Fig. 1. Model's feature extractor a) and classification module b).

Patch Importance Visualization: The extracted patches' relative importance can be computed by transforming the attention scores for the model's predicted class into percentiles and mapping them to their corresponding spatial location in the WSI. These scores are the output of the gated attention unit shown in Fig. 1. The result is presented in the form of an overlapping heatmap.

2.4 Self-supervised Contrastive Learning:

Contrastive learning schemes have being applied in previous skin cancer related studies to learn effective visual representations without the need for data labels [2,12,17]. In this work, we use Google's Simple framework for Contrastive Learning of visual Representations (SimCLR) [4]. This framework is composed of four major components:

- From a batch of size N, *data augmentation* is performed twice to generate, from each patch x, two patches $\tilde{x}_i$ and $\tilde{x}_j$. The result is a set of 2N patches with correlated views of the same examples. The data augmentations used are: *vertical and horizontal flip* with 50% probability each, *random cropping* with a resizing of the resulting image to the original size. *Color distortions* such as small changes in the image's hue or conversion to grayscale are applied with a probability of 80% and 20%, respectively. At the end of the data augmentation pipeline, all channel values were normalized using ImageNet's mean and variance.
- A *base encoder* $f(\cdot)$, the convolutional layers of a CNN, used to extract feature vectors h_i from each cropped patch $h_i = f(\tilde{x}_i)$.
- A *small projection head* $g(\cdot)$ in the form of an MLP with one hidden layer which maps h_i to z_i the space where contrastive loss is applied. That is, $z_i = g(h_i)$ where $z_i \in \mathbb{R}^j$, and $h_i \in \mathbb{R}^d$, and $j < d$.
- A *contrastive loss function* that encourages the clustering of z_i and z_j from the same original image together while separating them from the rest of the 2(N-1) patches.

The loss function is:

$$L_{i,j} = -\log \frac{\exp\left(\text{sim}\left(z_i, z_j\right)/\tau\right)}{\sum_{k=1}^{2N} \mathbb{K}_{[k \neq i]} \exp\left(\text{sim}\left(z_i, z_k\right)/\tau\right)} \qquad (2)$$

where $sim(z_i, z_j) = z_i^T z_j / \|z_i\| \|z_j\|$, measures the similarity between z_i and z_j. $\mathbb{K}_{[k \neq i]}$ is 1 if $k \neq i$, and 0 otherwise. τ denotes a temperature parameter. The total loss is computed across all pairs in a mini batch. A visual example can be found in Fig. 2, for a more detailed explanation, refer to [4].

3   Experimental Set-Up and Results

This section explains the two different sets of WSI features used and the experiments performed with them. Results are shown in Table 2 and Fig. 3.

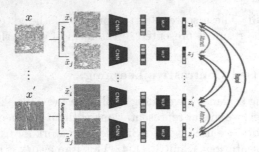

Fig. 2. Self-supervised contrastive learning framework. Each original patch x, is augmented twice and forwarded through a CNN. The features obtained after the projection head are compared using the cosine similarity. This loss function reinforces the clustering of vectors z_i and z_j while penalizing the similarity with the other $2(N-1)$ inputs.

3.1 Feature Extraction

Two ResNet50 [9] architectures were used to extract features from the cropped patches. Their pre-training scheme differed: the first CNN was pre-trained on ImageNet while the second was additionally fine-tuned using the SimCLR framework with a batch size of 64 patches. We denote the first set of features F_{Im} and the second F_{CLR}. The dimension of each embedding is $h_k \in \mathbb{R}^{2048}$.

3.2 Experiments

We trained six models to test the different methods presented in Sect. 2. We compare the standard, and the attention-based MIL assumptions explained in Sect. 2.2 with and without contrastive pre-training of the feature extractor. We also trained two more attention-based MIL models using clinical data. Additionally, we analyzed the relative importance given to each patch by the attention model when using F_{Im} versus F_{CLR}. The results are shown in Table 2 and Fig. 3 respectively.

We used a learning rate of 10^{-5} and L2 regularization (5×10^{-3}) with stochastic gradient descent as an optimizer. The specific values for learning rate, regularization, dropout probabilities, and the number of hidden layers for the clinical data input head and the classification MLPs were all selected using bayesian hyperparameter optimization. Cross entropy was used as a loss function when training the classifier. For testing, we chose the model weights with the best F1-score found in validation. Ten different random splits were used with a proportion of 80%, 10%, 10% for training, validation, and testing to obtain mean and variance.

4   Discussion

Table 2 shows that the standard MIL assumption did not yield better than random results in most computed metrics. The attention-based classifiers trained

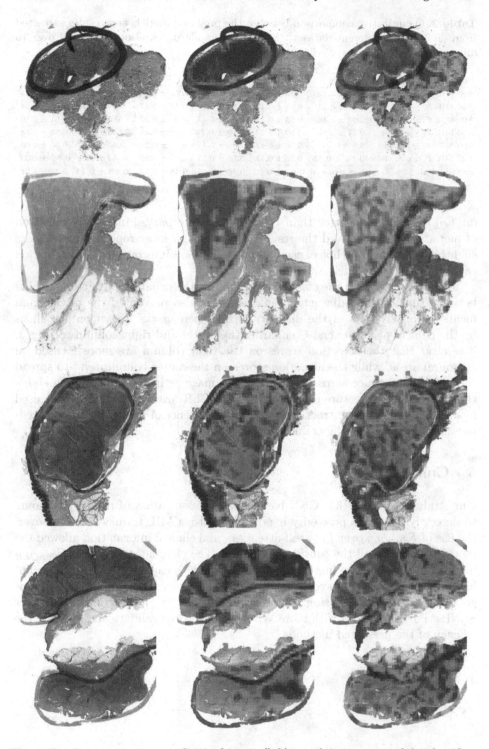

Fig. 3. Patch importance map. Original image (left), patch importance of the classifier trained on F_{Im} (center), and a second classifier trained on F_{CLR} (right).

Table 2. Quantitative comparison between the proposed models and results extracted from [3] (using a different dataset). Mean and standard deviation computed over 10 different splits.

Pre-training	F1-score	AUC	Bal. acc	Sensitivity	Specificity	Precision
Sta. MIL F_{Im}	0.51 ± 0.10	0.53 ± 0.13	0.50 ± 0.08	$\mathbf{0.71 \pm 0.26}$	0.29 ± 0.28	0.51 ± 0.22
Sta. MIL F_{CLR}	0.47 ± 0.11	0.53 ± 0.12	0.54 ± 0.10	0.49 ± 0.17	0.59 ± 0.18	0.50 ± 0.14
Att. MIL F_{Im}	0.52 ± 0.14	0.55 ± 0.12	0.55 ± 0.13	0.60 ± 0.22	0.5 ± 0.19	0.51 ± 0.10
Att. MIL F_{CLR}	0.57 ± 0.18	0.57 ± 0.23	$\mathbf{0.60 \pm 0.14}$	0.69 ± 0.26	0.513 ± 0.25	0.53 ± 0.18
Att. MIL F_{Im} & clinic	0.5 ± 0.14	0.56 ± 0.17	0.55 ± 0.13	0.67 ± 0.21	0.65 ± 0.27	0.50 ± 0.06
Att. MIL F_{CLR} & clinic	$\mathbf{0.58 \pm 0.15}$	0.58 ± 0.17	0.58 ± 0.13	0.65 ± 0.25	0.52 ± 0.28	$\mathbf{0.56 \pm 0.15}$
Brinker et al.	Not reported	$\mathbf{0.62 \pm 0.02}$	0.56 ± 0.02	0.48 ± 0.14	$\mathbf{0.65 \pm 0.11}$	0.44 ± 0.01

on F_{CLR} performed better than their F_{Im} counterpart. Finally, the addition of metadata also improved the results. Even though most reported metrics are slightly above random choice, each proposed method improved the classification capabilities of our model.

F_{CLR} also provided the classification module the capability of distinguishing between tumour and adjacent normal tissue. It also prevented the model from focalizing on artifacts of the dataset, such as pen marks drawn on the slides by the clinical practitioners. Comparing the center and right columns of Fig. 3, it is clear that the attention scores on the right column are more focused on tumoural areas, while the attention scores on the center column tend to spread the patch importance across the slide or on image artifacts. This finding shows that pre-training a feature extractor with SimCLR has the potential to be used for meaningful feature extractions even in the presence of artifacts without needing manual annotation by trained dermatologists.

5 Conclusions

Our study has shown that CNN-based image classification of primary tumours to detect lymph node positivity is possible under a MIL framework. Moreover, the use of F_{SimCLR} over F_{Im}, self-attention, and clinical information allowed the model to improve all the considered metrics. The classifier trained on F_{SimCLR} has shown a qualitative improvement in the model's capability to focus on clinically relevant areas and avoid image artifacts. However, the results are still not good enough to justify using deep learning systems as a selection method for SLNBs. Further studies with more WSIs are needed to validate and increase the efficacy of the presented methods.

References

1. Akdeniz, M., Hahnel, E., Ulrich, C., Blume-Peytavi, U., Kottner, J.: Prevalence and associated factors of skin cancer in aged nursing home residents: a multicenter prevalence study. PLoS ONE **14**(4), e0215379 (2019)
2. Barata, C., Santiago, C.: Improving the explainability of skin cancer diagnosis using CBIR. In: de Bruijne, M., et al. (eds.) MICCAI 2021. LNCS, vol. 12903, pp. 550–559. Springer, Cham (2021). https://doi.org/10.1007/978-3-030-87199-4_52
3. Brinker, T.J., Kiehl, L., Schmitt, M., et al.: Deep learning approach to predict sentinel lymph node status directly from routine histology of primary melanoma tumours. Eur. J. Cancer **154**, 227–234 (2021)
4. Chen, T., Kornblith, S., et al.: A simple framework for contrastive learning of visual representations. In: International Conference on Machine Learning, pp. 1597–1607. PMLR (2020)
5. Combalia, M., et al.: Validation of artificial intelligence prediction models for skin cancer diagnosis using dermoscopy images: the 2019 international skin imaging collaboration grand challenge. Lancet Digital Health **4**(5), e330–e339 (2022)
6. Gareau, D.S., et al.: Deep learning-level melanoma detection by interpretable machine learning and imaging biomarker cues. J. Biomed. Opt. **25**(11), 112906 (2020)
7. Gershenwald, J.E., et al.: Melanoma staging: evidence-based changes in the American joint committee on cancer eighth edition cancer staging manual. CA Cancer J. Clin. **67**(6), 472–492 (2017)
8. Goodfellow, I., Bengio, Y., Courville, A.: Deep Learning. MIT Press, Cambridge (2016)
9. He, K., Zhang, X., Ren, S., Sun, J.: Deep residual learning for image recognition. In: Proceedings of the IEEE Conference on Computer Vision and Pattern Recognition, pp. 770–778 (2016)
10. Ilse, M., Tomczak, J., Welling, M.: Attention-based deep multiple instance learning. In: International Conference on Machine Learning, pp. 2127–2136. PMLR (2018)
11. Jaiswal, A., Babu, A.R., Zadeh, M.Z., Banerjee, D., Makedon, F.: A survey on contrastive self-supervised learning. Technologies **9**(1), 2 (2020)
12. Li, X., Desrosiers, C., Liu, X.: Symmetric contrastive loss for out-of-distribution skin lesion detection. In: 2022 IEEE 19th International Symposium on Biomedical Imaging (ISBI), pp. 1–5. IEEE (2022)
13. Lu, M.Y., Williamson, D.F., et al.: Data-efficient and weakly supervised computational pathology on whole-slide images. Nat. Biomed. Eng. **5**(6), 555–570 (2021)
14. Otsu, N.: A threshold selection method from gray-level histograms. IEEE Trans. Syst. Man Cybern. **9**(1), 62–66 (1979)
15. Suganyadevi, S., Seethalakshmi, V., Balasamy, K.: A review on deep learning in medical image analysis. Int. J. Multimed. Inf. Retrieval **11**(1), 19–38 (2022)
16. Sung, H., et al.: Global cancer statistics 2020: GLOBOCAN estimates of incidence and mortality worldwide for 36 cancers in 185 countries. CA Cancer J. Clin. **71**(3), 209–249 (2021)
17. Verdelho, M.R., Barata, C.: On the impact of self-supervised learning in skin cancer diagnosis. In: 2022 IEEE 19th International Symposium on Biomedical Imaging (ISBI), pp. 1–5. IEEE (2022)

18. Vij, R., Arora, S.: Computer vision with deep learning techniques for neurodegenerative diseases analysis using neuroimaging: a survey. In: Khanna, A., Gupta, D., Bhattacharyya, S., Hassanien, A.E., Anand, S., Jaiswal, A. (eds.) International Conference on Innovative Computing and Communications. AISC, vol. 1388, pp. 179–189. Springer, Singapore (2022). https://doi.org/10.1007/978-981-16-2597-8_15
19. Wang, X., et al.: Weakly supervised deep learning for whole slide lung cancer image analysis. IEEE Trans. Cybern. **50**(9), 3950–3962 (2019)
20. Zhang, Z., et al.: Pathologist-level interpretable whole-slide cancer diagnosis with deep learning. Nat. Mach. Intell. **1**(5), 236–245 (2019)

Knowledge Distillation with a Class-Aware Loss for Endoscopic Disease Detection

Pedro E. Chavarrias-Solano[1]([✉]) [iD], Mansoor A. Teevno[1] [iD],
Gilberto Ochoa-Ruiz[1] [iD], and Sharib Ali[2] [iD]

[1] School of Engineering and Sciences, Tecnologico de Monterrey, Monterrey, Mexico
peterchavarrias@gmail.com
[2] School of Computing, University of Leeds, Leeds, UK

Abstract. Prevalence of gastrointestinal (GI) cancer is growing alarmingly every year leading to a substantial increase in the mortality rate. Endoscopic detection is providing crucial diagnostic support, however, subtle lesions in upper and lower GI are quite hard to detect and cause considerable missed detection. In this work, we leverage deep learning to develop a framework to improve the localization of difficult to detect lesions and minimize the missed detection rate. We propose an end to end student-teacher learning setup where class probabilities of a trained teacher model on one class with larger dataset are used to penalize multi-class student network. Our model achieves higher performance in terms of mean average precision (mAP) on both endoscopic disease detection (EDD2020) challenge and Kvasir-SEG datasets. Additionally, we show that using such learning paradigm, our model is generalizable to unseen test set giving higher APs for clinically crucial neoplastic and polyp categories.

Keywords: Deep learning · Object detection · Faster RCNN · Endoscopy disease detection · Knowledge distillation

1 Introduction

Gastrointestinal (GI) cancer accounts for 26% of the global cancer incidence and 35% of all cancer-related deaths, with colorectal, gastric and esophageal cancers being reported at 10.2%, 5.7% and 3.2% rates, respectively [3]. Clinical endoscopy is thus critical for disease detection, diagnosis and risk categorisation of patients, as it allows a visual interpretation of mucosal changes. However, the detection of subtle lesions such as dysplasia (early neoplasia) can be difficult, thus leading to 11.3% missed detection rates for neoplasia in the upper-GI and nearly 6% missed polyps in the lower-GI surveillance [17]. Artificial intelligence, and in particular, deep learning can help to improve the detection rates of hard to find lesions and minimise their missing rates, while assisting in their characterisation.

Although several methods of this type have been developed in the literature, most focus has been on the polyp class with many datasets being publicly released

© The Author(s), under exclusive license to Springer Nature Switzerland AG 2022
S. Ali et al. (Eds.): CaPTion 2022, LNCS 13581, pp. 67–76, 2022.
https://doi.org/10.1007/978-3-031-17979-2_7

and deep learning methods applied [16,18]. However, in reality, these methods cannot be used to find other inconspicuous lesions either on the same site or during a different endoscopic procedure. Positing this argument, the "Endoscopic disease detection challenge 2020 (EDD2020)" [1,2] released a dataset comprising of both upper-GI precancerous abnormalities such as Barrett's oesophagus, dysplasia, cancer and lower-GI anomalies including polyp and cancer. Motivated by this work, we aim to explore the opportunity this dataset has to offer to develop a unified deep learning framework for the entire GI tract. However, we also leverage other public datasets that have abundant polyp instances from the lower-GI surveillance (in our case Kvasir-SEG [10].

The main rationale for our work is related to the difficulty of distinguishing between polyps, neoplasia and NBDE lesions in the entirety of the GI tract (see Fig. 1). The EDD dataset is particularly challenging and it provides a good opportunity for testing methods that are robust and reliable in detecting such classes. As a matter of fact, the polyp class can be easily misclassified as neoplasia even for highly trained specialists, which can be problematic and thoroughly discussed above. Thus, herein we propose a method that leverages the teacher-student architecture used in knowledge distillation approaches as a way to mitigate these issues. Our approach makes use of a teacher network (a Faster RCNN object detector) trained on Kvasir-SEG [10] dataset to identify polyps with a higher certainty, whose layers are then frozen for the subsequent teaching process. In the next stage, this network is used to guide the training process of a student network (also a Faster RCNN) which is trained on a relatively small EDD2020 dataset [2] that contains polyps, neoplasia and NBDE lesion categories.

The rest of the paper is organized as follows. In Sect. 2, we discuss the state-of-the-art methods in endoscopic disease detection and the limitations of the existing approaches. In Sect. 3, we discuss how we use endoscopic datasets in our experiments and we introduce our knowledge distillation framework and the proposed class-aware penalization loss function. In Sect. 4, we present the experimental setup and provide qualitative and quantitative results. Finally, Sect. 5 concludes the paper.

2   Related Work

In the literature, several methods, including single-stage [8,18,20] and two-stage [16,21] networks, as well as anchor free architectures [19] have been used for the detection of different gastrointestinal diseases while maintaining real time performance. Some of the early approaches have exploited single stage detectors. For example, a SegNet [4] based model was proposed by [20] to detect polyps during colonoscopy procedures. A real time polyp detection pipeline in colonoscopy based on YOLO was developed by Uraban et al. [18]. In another approach [8], a single stage detector was leveraged for the early detection of multi-level esophageal cancer. Zhang et al. [22] used pre-trained ResYolo architecture to extract features and then optimised output through a convolutional

Table 1. Dataset sample distribution between training and test for EDD2020 (revised) and Kvasir-SEG

Dataset	Type	Train	Val	Test
EDD2020	Images	376	38	38
	NDBE	239	28	19
	Neoplasia	183	21	31
	Polyp	172	24	32
Kvasir-SEG	Images	800	100	100
	Polyp	858	111	102

Non-dysplastic Barrett's esophagus (NDBE)   Dysplasia (early neoplasia)   Polyp

Fig. 1. Representative samples of different classes in the EDD2020 dataset

tracker. An SSD (ssdgpnet) based model exploiting a feature pyramid network was proposed in [23] for gastric polyp detection. More recently, two-stage detectors have shown improved accuracy and robustness, e.g., methods reported in EDD(2020) [1]. A multi-stage detector uses a region proposal network to limit the search by generating candidates and then uses a classifier and bounding box regressor head to refine the search and produce final predictions. Among the two-stage networks, Faster RCNN [15] has been used as a base architecture in most detection pipelines owing to its improved precision. Yamada *et al.* [21] leveraged Faster RCNN with VGG as backone to detect hard samples of lesions normally ignored by colonoscopy procedures. They obtained encouraging results; however, the inference speed performance makes it unsuitable for real time examinations. In another study conducted by Shin *et al.* [16], Faster RCNN with Inception and ResNet backbones were used to detect polyps. The two stage detectors have performed better over their single stage counterparts in EDD(2020) challenge [1,12]. In another approach [11], a two stage framework based on feature pyramid prediction was proposed for polyp identification from colonoscopy images. Contrast enhanced colonoscopy images were fed to an improved Faster RCNN architecture to boost polyp detection performance by authors in [5]. Deep networks have been successful in yielding state-of-the-art results, but they require training huge models on large datasets for obtaining that performance. This makes them quite computationally expensive. An alternative approach is to use knowledge distillation for developing smaller and efficient models. This involves a trained teacher network to gradually transfer knowledge to a compact student network [7]. Leveraging this idea, a teacher-student mutual learning pipeline with single teacher, multi-student framework was proposed in [14] for the polyp detection using hyper-Kvasir dataset. Henrik *et al.* [6] used a knowledge distillation framework to perform a semi-supervised polyp classification showing a few points improvement in dice score over previous methods. Self-distillation was used for the early diagnosis of neoplasia in Barrett's esophagus [9]. Knowledge distillation along with video temporal context was used to implement a real-time polyp detection framework by Li *et al.* [13]. Motivated by these works, we propose to use a fully trained (teacher) network on one class and use the predicted class probability to penalise the multi-class (student) learner.

3   Materials and Method

3.1   Datasets

The EDD2020 dataset [1] includes multi-center, multi-modal, multi-organ, multi-disease images comprising of both upper and lower GI. The dataset is encompassed by 386 video frames with 749 annotations that were collected using three different endoscopic modalities (white light, narrow-band imaging, and chromoendoscopy) at four different clinical centers during the screening of four different gastrointestinal organs. The detection and segmentation task is related to five disease categories: non-dysplastic Barrett's (286), suspicious (98), high-grade dysplasia (80), cancer (57), and polyp (228). Barrett's (NDBE) and polyp classes have more recurrences, while cancer, high-grade dysplasia (HGD), and suspicious have less samples. Some representative samples of the dataset are shown in Fig. 1. EDD2020 dataset has large class imbalance so we merged cancer, HGD and suspicious classes into one class as "neoplasia". The final data distribution is shown in Table 1. We have made use of an additional dataset, the Kvasir-SEG [10] that is widely used to develop and compare methods built for polyp detection and segmentation in the colon and rectum. This dataset contains 1,000 polyp images with their corresponding bounding boxes and segmentation masks. The second row in Table 1 also summarizes the sample sizes of this dataset.

3.2   Proposed Knowledge-Distillation Framework

Figure 2 represents the block diagram of our proposed network. Here, the class probability distribution vectors of both networks are used to compute the Bhattacharyya distance for each class, which is then used as a penalization term together with the standard cross-entropy loss for optimizing the parameters of the student model. The components of our framework are described as follows:

Faster R-CNN an unified network for object detection that is composed of two main modules: a deep fully convolutional network that proposes regions (*aka* Region Proposal Network (RPN)), and the Fast R-CNN detector that uses the pooled areas from the proposed regions [15] for classification and bounding box regression. The RPN takes the output of the last shared convolutional layer as input and produces a set of rectangular object proposals, each with an objectness score. To generate region proposals, the RPN takes as input a $n \times n$ window of the input feature map and maps it to a lower-dimensional feature. It is then fed into two fully-connected layers: a box-regression layer and a box-classification layer. At each window, the algorithm predicts multiple region proposals which are parameterized relative to the reference boxes, known as anchors centered at their corresponding window. This approach allows sharing features without extra computations for capturing different size objects in a multi-scale fashion [15].

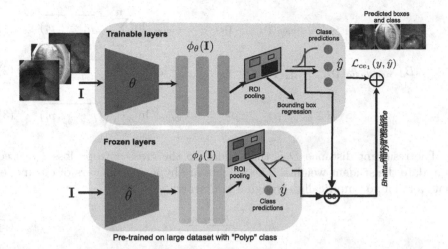

Fig. 2. Featured-based knowledge distillation (proposed framework). The student (trainable layers) and teacher (frozen layers) networks are both implemented using a Faster R-CNN architecture. Here, the teacher model layers are pre-trained on a larger polyp dataset after which they are frozen. The feature vectors of class wise probability distributions are extracted from intermediate layers of both models. The Bhattacharyya distance is then computed between these probability distributions of both networks for each class separately, with varying weights and is added as a penalization term to the student classification loss.

Class-Aware Loss Function. We propose a class-aware loss to boost the performance of the object detection model. Our implementation follows a feature-based knowledge distillation approach [7], in which we use a Faster R-CNN teacher network pre-trained in the Kvasir-SEG dataset for detecting polyps with a high confidence. The student network, also a Faster R-CNN architecture, compares the probability maps generated by its feature extraction network and those of the teacher network class probabilities. The rationale for this process is to implement a manner for penalizing the student network each time a polyp sample is classified incorrectly by the student network, but correctly detected by the teacher. Severe neoplastic cases that can confuse with polyp class are penalized with higher weights as that those with polyps. Similarly, for the NDBE cases, the penalisation factor (λ) is set relatively lower as these are slightly different disease cases but can confuse with early neoplastic changes. These λ weights for each disease cases were empirically set.

Bhattacharyya distance (B_d) is computed between each prediction logits of the student (p_s) and teacher networks (p_t), as shown in Eqs. (1–3). A set of experimentally determined fixed weighting factors λ are assigned to each distance. The sum of all distances is then computed and normalized as in Eq. (4). The normalization factor is determined by adding weights determined for each class ($\sum \lambda_{class} = \lambda_{ndbe} + \lambda_{neoplasia} + \lambda_{polyp}$).

$$B_{d-ndbe}(\mathbf{p}_{s_{ndbe}}, \mathbf{p}_t) = -\lambda_{ndbe} \cdot \ln(\sum_{i=1}^{n} \sqrt{p_{s_{ndbe}}^i, p_t^i}) \quad (1)$$

$$B_{d-neoplasia}(\mathbf{p}_{s_{neoplasia}}, \mathbf{p}_t) = -\lambda_{neoplasia} \cdot \ln(\sum_{i=1}^{n} \sqrt{p_{s_{neoplasia}}^i, p_t^i}) \quad (2)$$

$$B_{d-polyp}(\mathbf{p}_{s_{polyp}}, \mathbf{p}_t) = -\lambda_{polyp} \cdot \ln(\sum_{i=1}^{n} \sqrt{p_{s_{polyp}}^i, p_t^i}) \quad (3)$$

The resultant distance D is then added to the cross-entropy loss $\mathcal{L}_{ce}$ used to update the student weights. The weights are hyper-parameters of our model that were found empirically through a grid search.

$$D = \frac{\sum B_{d-class}}{\sum \lambda_{class}} \quad (4)$$

Data Augmentation Techniques. Two types of data augmentation techniques were applied to both datasets: geometric and photometric. For the geometric data augmentations we refer to 90° random rotation, horizontal flip, vertical flip, and center crop. While, photometric image augmentations entail to transformations such as blur, image equalization, as well as changes in contrast, brightness, darkness, sharpness, hue, and saturation of the images. All augmentations are only applied to training set.

4 Experiments and Results

4.1 Experimental Setup and Evaluation Metrics

All models were trained on an NVIDIA GeForce GTX 1650 GPU. Images were resized to 225×225 pixels and the following training, validation and testing splits is performed on both datasets with an 80:10:10 split. The data augmentation techniques described in Sect. 3.2 are used in all our models and we use k-fold with $k = 3$ cross validation for unbiased comparison between models. All the Faster R-CNN sub-modules in our teacher-student framework make use of a ResNet-50 backbone and were run with similar hyper-parameters that include a learning rate of 1e−3, momentum of 0.9, and a weight decay of 5e−4. Stochastic Gradient Descent (SGD) was used as an optimizer.

We use standard computer vision metrics for detection that includes average precision (AP) computed for each class and mean average precision (mAP) over all three classes. Each of these metrics are evaluated at different IoU thresholds and are usually represented as $\text{AP}_{\text{IoU threshold}}$. We present results for IoU thresholds of 25, 50 and 25:75 (meaning averaged value for IoU threshold between 25 and 75 with spacing of 5). Usually, AP_{50} is considered to have indicative results with good localization and classification score.

Table 2. Evaluation on Kvasir-SEG dataset using Faster R-CNN. Evaluation results using mAP_{25}, mAP_{50}, mAP_{75}, and $mAP_{25:75}$ evaluation metrics

Method	Augmentation	Epochs	mAP_{25}	mAP_{50}	mAP_{75}	$mAP_{25:75}$
Faster R-CNN	None	60	95.3	90.4	68.8	86.6
(ResNet-50)	3-Fold Cross Validation done at mAP_{50}					
			K1	K2	K3	Average
	Geometric	20	82.4	85.5	85.1	84.3
	Photometric	20	87.3	83.6	86.5	85.8
	Geometric+photometric	20	**87.6**	**85.6**	**88.1**	**87.1**

Table 3. Evaluation on 10% held-out test data. Faster R-CNN (ResNet-50) without (SOTA) and with informed polyp category knowledge distillation (proposed) for which class aware weights are $\lambda_{ndbe} = 0.165$, $\lambda_{neoplasia} = 0.33$, $\lambda_{polyp} = 0.33$

Method	Augment	Epochs	mAP_{25}	mAP_{50}	$mAP_{25:75}$
Faster R-CNN (ResNet-50)	None	60	50.2	40.3	**37.9**
Proposed	None	60	**53.9**	**42.1**	37.8
			Class wise average precision		
			$AP_{50_{ndbe}}$	$AP_{50_{neoplasia}}$	$AP_{50_{polyp}}$
Faster R-CNN (ResNet-50)	None	60	65.6	**42.2**	12.9
Proposed	None	60	**71.9**	37.3	**16.9**

4.2 Results

We first trained Faster R-CNN using the Kvasir-SEG dataset. Then, we applied an informed class-aware knowledge distillation to the student network by applying the frozen pre-trained model on Kvasir-SEG (larger dataset on with polyp class) to improve the performance of the model on the EDD2020 dataset by leveraging the class-aware knowledge of this network.

Quantitative Results. Table 2 present the results on Kvasir-SEG test split showing an improvement (nearly 3% more compared to other augmentation) with our proposed combined geometric and photometric augmentation strategy. Table 3 demonstrate that the use of class-aware loss in our proposed knowledge distillation concept improves the mAP at different IoU thresholds for hold-out test samples (3.7% on mAP_{25} and 2.2% on mAP_{50}). Similarly, for the class wise APs, a boost in NDBE and polyp classes can be observed with over 6% and 4%, respectively. Similarly, Table 4 demonstrate that the proposed approach outperformed the Faster R-CNN approach at all mAPs. Also, for the neoplasia class and polyp class, our approach showed an improvement of 3.3% and nearly 13%, respectively.

74   P. E. Chavarrias-Solano et al.

Table 4. Evaluation on unseen test data. Faster R-CNN (ResNet50) without (SOTA) and with informed polyp class knowledge distillation (proposed, with class aware loss weights $\lambda_{ndbe} = 0.165$, $\lambda_{neoplasia} = 0.33$, $\lambda_{polyp} = 0.33$).

Method	Augmentation	Epochs	mAP$_{25}$	mAP$_{50}$	mAP$_{25:75}$
Faster R-CNN (ResNet-50)	None	60	46.6	38.9	34.6
Proposed	None	60	**48.9**	**42.6**	**36.9**
			Class wise average precision		
			AP50_{ndbe}	AP$50_{neoplasia}$	AP50_{polyp}
Faster R-CNN (ResNet-50)	None	60	**47.9**	17.8	51.0
Proposed	None	60	42.8	**21.1**	**63.7**

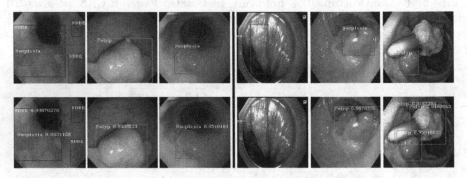

Fig. 3. Qualitative results. First row shows images with their corresponding bounding boxes. Second row contains their corresponding predictions. (Color figure online)

Qualitative Results. Figure 3 presents a qualitative results on six samples of test set with their corresponding ground truth bounding boxes (on top) and detected boxes with classes from our proposed approach (on bottom). It can be observed that while protruded polyp class, semi-protruded neoplasia are detected precisely due to their appearance, however, there is a confusion with similar looking neoplastic changes (e.g., 5th column) or even other visual clutters (e.g., 6th column). Also, due to heterogeneous samples in neoplasia class (refer 4th column), the algorithm fails on under represented samples. It is also clear that in presence of multi-class (see 1st column), the algorithm is able to detect both non-dysplastic areas (in blue boxes) and early neoplasia (in red box).

5 Conclusion

We have presented a class-aware knowledge distillation approach by leveraging class probability distances. The teacher branch is first trained using a larger dataset that is used to penalize the performance of the student model. Here, we proposed to use Bhattacharyya distance between class probability distributions. We demonstrate that our approach improves overall detection accuracies and can generalize well with the unseen test set, in particular, neoplasia and polyp that are clinically more important than non-dysplastic areas. In future work, we will use a third teacher model trained on another unique class.

Acknowledgments. The authors wish to thank the AI Hub and the CIIOT at ITESM for their support for carrying the experiments reported in this paper in their NVIDIA's DGX computer.

References

1. Ali, S., et al.: Deep learning for detection and segmentation of artefact and disease instances in gastrointestinal endoscopy. Med. Image Anal. **70**, 102002 (2021). https://doi.org/10.1016/j.media.2021.102002. arXiv: 2010.06034
2. Ali, S., et al.: Endoscopy disease detection challenge 2020. arXiv preprint arXiv:2003.03376 (2020)
3. Arnold, M., et al.: Global burden of 5 major types of gastrointestinal cancer. Gastroenterology **159**(1), 335-349.e15 (2020)
4. Badrinarayanan, V., Kendall, A., Cipolla, R.: SegNet: a deep convolutional encoder-decoder architecture for image segmentation. IEEE Trans. Pattern Anal. Mach. Intell. **39**(12), 2481–2495 (2017)
5. Chen, B.L., Wan, J.J., Chen, T.Y., Yu, Y.T., Ji, M.: A self-attention based faster R-CNN for polyp detection from colonoscopy images. Biomed. Signal Process. Control **70**, 103019 (2021)
6. Gjestang, H.L., Hicks, S.A., Thambawita, V., Halvorsen, P., Riegler, M.A.: A self-learning teacher-student framework for gastrointestinal image classification. In: 2021 IEEE 34th International Symposium on Computer-Based Medical Systems (CBMS), pp. 539–544. IEEE (2021)
7. Gou, J., Yu, B., Maybank, S.J., Tao, D.: Knowledge distillation: a survey (2020). https://doi.org/10.48550/ARXIV.2006.05525. https://arxiv.org/abs/2006.05525
8. Horie, Y., et al.: Diagnostic outcomes of esophageal cancer by artificial intelligence using convolutional neural networks. Gastrointest. Endosc. **89**(1), 25–32 (2019)
9. Hou, W., Wang, L., Cai, S., Lin, Z., Yu, R., Qin, J.: Early neoplasia identification in Barrett's esophagus via attentive hierarchical aggregation and self-distillation. Med. Image Anal. **72**, 102092 (2021)
10. Jha, D., et al.: Kvasir-SEG: a segmented polyp dataset. In: Ro, Y.M., et al. (eds.) MMM 2020. LNCS, vol. 11962, pp. 451–462. Springer, Cham (2020). https://doi.org/10.1007/978-3-030-37734-2_37
11. Jia, X., et al.: Automatic polyp recognition in colonoscopy images using deep learning and two-stage pyramidal feature prediction. IEEE Trans. Autom. Sci. Eng. **17**(3), 1570–1584 (2020)
12. Krenzer, A., Hekalo, A., Puppe, F.: Endoscopic detection and segmentation of gastroenterological diseases with deep convolutional neural networks. In: EndoCV@ ISBI, pp. 58–63 (2020)
13. Li, X., Liu, R., Li, M., Liu, Y., Jiang, L., Zhou, C.: Real-time polyp detection for colonoscopy video on CPU. In: 2020 IEEE 32nd International Conference on Tools with Artificial Intelligence (ICTAI), pp. 890–897. IEEE (2020)
14. Niyaz, U., Bathula, D.R.: Augmenting knowledge distillation with peer-to-peer mutual learning for model compression. In: 2022 IEEE 19th International Symposium on Biomedical Imaging (ISBI), pp. 1–4. IEEE (2022)

15. Ren, S., He, K., Girshick, R., Sun, J.: Faster R-CNN: towards real-time object detection with region proposal networks. arXiv:1506.01497, January 2016. http://arxiv.org/abs/1506.01497
16. Shin, Y., Qadir, H.A., Aabakken, L., Bergsland, J., Balasingham, I.: Automatic colon polyp detection using region based deep CNN and post learning approaches. IEEE Access **6**, 40950–40962 (2018)
17. Turshudzhyan, A., Rezaizadeh, H., Tadros, M.: Lessons learned: preventable misses and near-misses of endoscopic procedures. World J. Gastrointest. Endosc. **14**(5), 302–310 (2022)
18. Urban, G., et al.: Deep learning localizes and identifies polyps in real time with 96% accuracy in screening colonoscopy. Gastroenterology **155**(4), 1069–1078 (2018)
19. Wang, D., et al.: AFP-Net: realtime anchor-free polyp detection in colonoscopy. In: 2019 IEEE 31st International Conference on Tools with Artificial Intelligence (ICTAI), pp. 636–643. IEEE (2019)
20. Wang, P., et al.: Development and validation of a deep-learning algorithm for the detection of polyps during colonoscopy. Nat. Biomed. Eng. **2**(10), 741–748 (2018)
21. Yamada, M., et al.: Development of a real-time endoscopic image diagnosis support system using deep learning technology in colonoscopy. Sci. Rep. **9**(1), 1–9 (2019)
22. Zhang, R., Zheng, Y., Poon, C.C., Shen, D., Lau, J.Y.: Polyp detection during colonoscopy using a regression-based convolutional neural network with a tracker. Pattern Recogn. **83**, 209–219 (2018)
23. Zhang, X., et al.: Real-time gastric polyp detection using convolutional neural networks. PLoS ONE **14**(3), e0214133 (2019)

IF³: An Interpretable Feature Fusion Framework for Lesion Risk Assessment Based on Auto-constructed Fuzzy Cognitive Maps

Georgia Sovatzidi[ID], Michael D. Vasilakakis[ID], and Dimitris K. Iakovidis[✉][ID]

University of Thessaly, Papasiopoulou Street 2-4, 35131 Lamia, Greece
{gsovatzidi,vasilaka,diakovidis}@uth.gr

Abstract. The detection of abnormalities in the gastrointestinal (GI) tract, including precancerous lesions, is substantially subject to expert knowledge and experience. To address the challenge of automated lesion risk assessment, based on Wireless Capsule Endoscopy (WCE) images, this paper introduces a novel Artificial Intelligence (AI) framework based on Fuzzy Cognitive Maps (FCMs). Specifically, FCMs are fuzzy graph structures used to model knowledge spaces using cause-and-effect relationships, enabling uncertainty-aware reasoning and inference. The novel proposed Interpretable FCM-based Feature Fusion (IF³) framework, includes the following contributions: a) it automatically constructs an FCM based on similarities discovered in training data; b) it enables the fusion of different features extracted using different methods. The proposed framework is generic, domain-independent and it can be integrated into any classifier. To demonstrate its performance, experiments were conducted using real datasets, which include a variety of GI abnormalities, and different feature extractors. The results show that the automatically constructed FCM outperforms state-of-the-art methods, while providing interpretable results, in an easily understandable way.

Keywords: Gastrointestinal tract · Precancerous lesions · Fuzzy Cognitive Maps · Feature Fusion · Fuzzy sets · Interpretability

1 Introduction

Cancer remains one of the leading causes of mortality, as it is responsible for one in six deaths worldwide, according to the World Health Organization (WHO). Early detection of cancerous or precancerous lesions is an important factor in reducing cancer mortality. Representative examples of precancerous lesions in the gastrointestinal (GI) tract are polyps and ulcers [1]. Although polyps and ulcers of the GI tract are mainly benign, there are cases that can be evolved into malignant tumors [2].

In the last decade, many efforts have been devoted to the development of AI models aiming to perform the examination of the GI tract [3, 4] and the early detection of cancer [5]. Recently proposed CNN models include the Look-Behind Fully CNN (LB-FCN) [6], and the Weakly Supervised CNN (WCNN) [7], which were proposed for the automatic

© The Author(s), under exclusive license to Springer Nature Switzerland AG 2022
S. Ali et al. (Eds.): CaPTion 2022, LNCS 13581, pp. 77–86, 2022.
https://doi.org/10.1007/978-3-031-17979-2_8

detection and localization of GI abnormalities. In [8] a polyp detection approach was proposed, based on the Scale Invariant Feature Transform (SIFT) algorithm and the Complete Local Binary Pattern (CLBP). Although there has been a significant progress in explaining the decisions of machine learning models, their adoption is limited, due to their "black-box" nature [9]. Specifically, many of these models still lack of providing justifications with respect to their predictions [10], thus limiting user trust. Recently, a state-of-the-art feature extraction model, named Explainable Fuzzy Bag of Words (XFBoW), was introduced to perform interpretable endoscopic image analysis [11].

The ability of an AI model to provide an easily understandable interpretation of its prediction to a user, could be considered of similar importance to a physician justifying accurately his diagnosis to a patient. There are many benefits in using structured knowledge in the form of graphs to generate explanations, such as high understandability, reactivity, and accuracy [12]. FCMs are fuzzy-graph structures mainly used for representing causal reasoning [13]. Due to their effectiveness and simplicity in implementation, FCMs have been widely used in many applications of different scientific fields [14]. However, a limitation of such graph-based models is that there is a need of human intervention, as experts with certain knowledge of the system being modelled, need to participate for their determination. To overcome this limitation, this paper introduces a novel Interpretable FCM-based Feature Fusion (IF^3) framework that includes the following contributions: a) it automatically constructs an FCM based on similarities discovered in training data; b) it enables the fusion of different features extracted using different methods. The proposed framework is generic, domain-independent and can be integrated into any classifier, including CNNs, to support the resulting decision, providing easily understandable explanations. The IF^3 can be considered as an extension of XFBoW, with the aim of further improving the interpretability of the model and is developed in a way that is understandable from the perspective of users, without requiring any prior knowledge. The rest of this paper is organized as follows: Sect. 2 presents the proposed framework. Experiments and results are presented in Sect. 3, and, in Sect. 4 conclusions derived from this study, are summarized.

2 Methodology

2.1 Fuzzy Cognitive Maps

Fuzzy Cognitive Maps are signed directed graphs consisting of nodes and weighted arcs. Nodes of the graph stand for the concepts $C = C_1, C_2, ..., C_n$ where n is the number of concepts, describing behavioral characteristics of the examined system. The relationship between two concepts, is expressed with the notion of a weight w_{ij}, $i = 1, ..., C_n, j = 1, ..., C_n$ that can take a value in the interval $[-1, 1]$. There are three possible types of interaction: negative ($w_{ij} < 0$), positive ($w_{ij} > 0$), and neutral ($w_{ij} = 0$) [13]. For the definition of the structure of an FCM, experts evaluate and describe the number of concepts needed for a given problem along with the existing relations among them. After the determination of the model, the reasoning phase takes place. Let $A^t = \left(a_1^t, a_2^t, a_3^t, ..., a_n^t\right)$ be the activation vector produced by an FCM. The reasoning procedure takes place, using an activation transfer function, according to Eq. 1, until the FCM finally converges to a steady state:

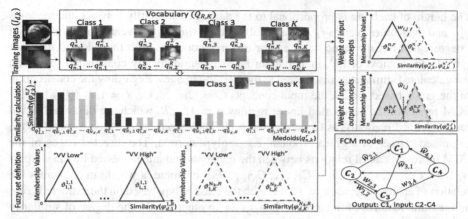

Fig. 1. Definition of FCM-weights based on the proposed IF^3 framework.

$$a_i^{t+1} = \sigma\left(a_i^t + \sum_{j=1, j \neq i}^{n} w_{ji} a_j^t\right) \tag{1}$$

where w_{ji} is the weight that expresses the causal relation connecting C_j to C_i, σ is a transfer function, and a_j^t denotes the activation value of a concept at time t. The initial activation state vector A^0 represents the initial state vector.

2.2 Proposed Framework

To automatically construct FCMs based on the introduced IF^3 framework, the following four steps have to be performed. In step 1, the feature extraction procedure based on the XFBoW model precedes [11]. Let us consider a set of training images $I_{d,k}, d = 1, \ldots, D_k, k = 1, \ldots, K$ where K represents the number of different classes of a given problem. To extract features from the training images, "dense sampling" is used [11], and segment each image into R regions (Fig. 1). From each region a set of feature vectors $f_{d,r}, r = 1, .., R$, such as color and textural features, are extracted. The feature vectors are then clustered into N_k groups with medoids $q_{n,k}^r, n = 1, \ldots, N_k, r = 1, .., R$, which correspond to representative image regions of each class. The generated medoids constitute a vocabulary $Q_{R,K} = \left\{\{q_{n,k}^1\}, \ldots, \{q_{n,k}^R\}\right\}$.

In step 2, the visual content of the training images is described, based on the generated vocabulary. Specifically, the similarity $\varphi_{d,k}^{n,r}$ among the extracted feature vectors $(f_{d,r})$ and the generated medoids $(q_{n,k}^r)$ is calculated [11]. To interpret the calculated similarities in an easily understandable way, linguistic values are utilized. Thus, in step 3, a number of fuzzy sets $\vartheta_{l,k}^{n,r}, l = 1 \ldots L$ is defined, based on a corresponding triangular membership function $\mu_{l,k}^{n,r}\left(\varphi_{d,k}^{n,r}\right)$ [11, 15]. Indicative examples of such linguistic values are "Very Low", "Medium", and "Very High", which are utilized to characterize the calculated similarities $\varphi_{d,k}^{n,r}$. In Fig. 1, the calculated similarities are depicted with colored bars.

The height of each bar is proportional to the degree of similarity between the medoids $q^r_{n,k}$ and the feature vectors $f_{d,r}$. Specifically, a high similarity reveals that the examined image region has the same color and/or texture properties with the selected medoids [16]. In the context of IF^3, in step 4, an FCM is automatically constructed, based on the calculated similarities discovered in the training images. Regarding the construction of the graph-based model, the input concepts $C = C_1, \ldots, C_s, s = 1, .., |Q_{R,K}|$ of the FCM represent the segmented image regions $r = 1, .., R$, which are described based on the defined medoids, as $C_s = q^r_{n,k}$ (step 1). The output concepts correspond to the examined classes of a given problem, e.g., polyp, normal. The directed edges of the FCM represent causal relations between the concepts and are expressed by the weights $w_{ij}, i = C_1, \ldots, C_{|Q_{R,K}|}, j = C_1, \ldots, C_{|Q_{R,K}|}$. For example, a weight may express the relation of an image region $r = 1, .., R$ with the class polyp, based on the extracted color and textural features of the training images; or in other words, the degree of similarity $(\varphi^{n,r}_{d,k})$ that exist among the r and the respective classes.

During the construction of the conventional FCMs, medical experts gather, and their different opinions are aggregated aiming to describe the relations between the concepts. Based on this procedure, the determination of the weights of the graph is performed by aggregating the defined fuzzy sets of step 3, using the fuzzy union and intersection operations [15]. The union operation aggregates the fuzzy sets $\vartheta^{n,r}_{l,k}$ that describe the similarities among all the $f_{d,r}$ to the output concepts $C_k, k = 1, \ldots, K$ and expressed as $\vartheta^{n,r}_k = \bigcup^L_{l=1} \vartheta^{n,r}_{l,k}$ with membership functions $\mu^{n,r}_k = \max \mu^{n,r}_{l,k}$. The intersection operation is utilized to fuse the fuzzy sets $\vartheta^{n,r}_k, \vartheta^{n',r'}_k$ that describe the similarities among all the $f_{d,r}, f_{d,r'}$ to the input concepts $C_r, C_{r'}$, where $n, n' = 1 \ldots N_k$ and $n \neq n', r, r' = 1 \ldots R$ and $r \neq r'$. The intersection results to the fuzzy set $\vartheta^{n,n'}_{r,r'} = \vartheta^{n,r}_k \cap \vartheta^{n',r'}_k$ with membership functions $\mu^{n,n'}_{r,r'} = \min(\mu^{n,r}_k, \mu^{n',r'}_k)$. Two types of weights can be distinguished: the weight existing between: a) input concepts that are described by the fuzzy sets $\vartheta^{n,n'}_{r,r'}$ (Eq. 2), and b) an input and an output concept, described by the fuzzy sets $\vartheta^{n,r}_k$ (Eq. 3) The interconnection between two input concepts is given by Eq. (2).

$$w_{i,j} = \sum_{k,k'} \sum_{n,n'} \sum^{D_k}_{d=1} \varphi^{n,r}_{d,k} * \mu^{n,n'}_{r,r'}(\varphi^{n,r}_{d,k}) / \sum_{k,k'} \sum_{n,n'} \sum^{D_k}_{d=1} \mu^{n,n'}_{r,r'}(\varphi^{n,r}_{d,k}) \quad (2)$$

where $k', k = 1 \ldots K$ and $n, = 1, \ldots, N_k, n' = 1, \ldots, N_{k'}$, with $n \neq n'$ with r, r' indicate the respective concepts $C_s = q^r_{n,k}, C_{s'} = q^{r'}_{n',k'}$ The interconnection between input and output concepts is calculated as follows:

$$\widehat{w}_{i,j} = \sum^{D_k}_{d=1} \varphi^{n,r}_{d,k} * \mu^{n,r}_k(\varphi^{n,r}_{d,k}) / \sum^{D_k}_{d=1} \mu^{n,r}_k(\varphi^{n,r}_{d,k}) \quad (3)$$

where $k = 1 \ldots K$ and, $n = 1, \ldots, N_k$ with r indicate the respective input concept C_s to an output concept C_k. Given an initial state vector A^0, the concept node values $(a^0_{r,k})$ of the FCM are calculated based on Eq. (4). In addition, the FCM starts the reasoning process, when $A^0 = (a^0_{1,1}, \ldots, a^0_{R,K})$ is inserted to the model and it proceeds until it reaches a convergence state.

$$a^0_{r,k} = \mu^{n,r}_{l,k}\left(\varphi^{n,r}_{d,k}\right) \cdot \underset{l=1}{\overset{L}{argmax}}\left(\mu^{n,r}_{l,k}\left(\varphi^{n,r}_{d,k}\right)\right) / L \quad (4)$$

where $r = 1, \dots R$, $\varphi_{d,k}^{n,r}$ is the similarity of a feature vector with the medoids $q_{n,k}^r$. The variable $\mu_{l,k}^{n,r}$ is the value of the membership function, $argmax_{l=1}^{L}\left(\mu_{l,k}^{n,r}\left(\varphi_{d,k}^{n,r}\right)\right)$ represents the fuzzy set with the maximum calculated membership value, and L corresponds to the defined fuzzy sets.

3 Experiments and Results

3.1 Dataset Description and Parameter Settings

Experiments were conducted using the publicly available: a) KID Dataset 2 [17], where the included WCE images are 303 with vascular anomalies, 44 polypoid, 227 inflammatory, and 1778 normal. b) A subset of 6000 of the Kvasir dataset [18] that contains labelled images of the classes: ulcer, polyp, blood, angiectacia and normal mucosa. For the vocabulary construction, the k-means algorithm [19] was used. The vocabulary size ranges between 20 to 250 medoids. The images were sampled in 18×18 pixel regions for the feature extraction [20]. The color features are formed by the *CIE*-Lab values (L, a, b) and the minimum and maximum values of all of the three components, within a square region [11]. The components of *CIE*-Lab represent the quantity of red ($a > 0$), green ($-a > 0$), as well as yellow ($b > 0$), and blue ($-b > 0$) of a pixel [16]. The textural features are extracted, using a 1-level 2D Discrete Wavelet Transform (DWT) on the L component of the *CIE*-Lab image representation. In addition, a 10-fold cross validation evaluation scheme was adopted. All the experiments were executed for 15 independent runs on a workstation with an Intel i5 2.5 GHz CPU and 4 GB RAM. For the visualization of the graph, the "FCM expert" tool was utilized [21].

3.2 Interpretable Example of Risk Assessment Using IF^3

To better understand how IF^3 provides interpretable results, an illustrative example from a fold of the 10-fold cross validation performed, is provided, aiming to assess the risk of the presence of a lesion in the GI tract. Let us consider an unknown input testing image, the visual content of which needs to be identified. The generated vocabulary $Q_{4,4}$ (Fig. 2) consists of the following four classes, each of which includes the corresponding medoids: *"normal"* ($q_{1,1}^1$ to $q_{1,1}^4$), *"inflammatory (infl.)"* ($q_{1,2}^1$ to $q_{1,2}^4$), *"polyp"* $\left(q_{1,3}^1$ to $q_{1,3}^4\right)$, *"vascular (vasc.)"* ($q_{1,4}^1$ to $q_{1,4}^4$). To perform the lesion risk assessment based on IF^3, the feature vectors (f_r) with the examined features, *i.e.*, color and texture, are extracted from the input image. After following the steps described in the Subsect. 2.2, the similarities among the f_r and the corresponding medoids $\left(q_{1,1}^1$ to $q_{1,4}^4\right)$ are calculated. In the sequel, the similarities are characterized linguistically, and they are fuzzified using the following seven linguistic terms. The respective similarity value intervals are provided in parentheses, which are in overlapping intervals meaning that a data vector may belong partially to more than one fuzzy sets [22]: "Very Very Low (VVL)" [0, 0.25], "Very Low (VL)" [0.125, 0.375], "Low (L)" [0.25, 0.5], "Medium (M)" [0.375, 0.625], "High (H)" [0.5, 0.75], "Very High (VH)" [0.625 − 0.875], "Very Very High (VVH)" [0.75 − 1].

With respect to the problem under consideration, the FCM constructed, using the proposed IF^3 framework, is illustrated in Fig. 2. The generated FCM consists of sixteen input concepts, *i.e.*, $C_1 - C_{16}$, which correspond to the most representative segmented image regions $r = 1, .., 4$ per examined class and they are described based on the defined medoids $q_{1,1}^1 - q_{1,4}^4$. The output concepts of the FCM are $C_{17} - C_{20}$ and their corresponding values are represented by the four last parameters of A^0, *i.e.*, *normal, inflammatory, vascular, polyp*. In the graph, the weights express the relation of an image region $r = 1, .., R$ with the examined classes, *i.e.*, *normal, polyp, inflammatory, vascular* based on the extracted color and textural features of the training images; or in other words, the calculated degree of similarity ($\varphi_{d,k}^{n,r}$) that exist among the examined image regions and the respective classes. Thus, for the FCM of the examined problem, the values of the initial state vector A^0 are calculated based on Eq. (4):

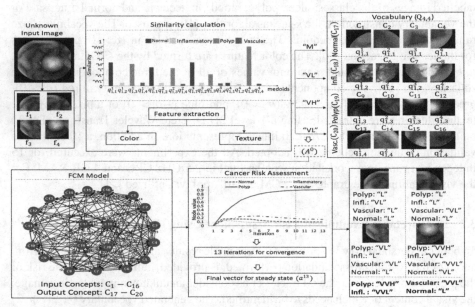

Fig. 2. Interpretable example of lesion risk assessment using IF^3.

$$A^0 = \begin{bmatrix} q_{1,1}^1, q_{1,1}^2, q_{1,1}^3, q_{1,1}^4, q_{1,2}^1, q_{1,2}^2, q_{1,2}^3, q_{1,2}^4, q_{1,3}^1, q_{1,3}^2, q_{1,3}^3, q_{1,3}^4, q_{1,4}^1, q_{1,4}^2, q_{1,4}^3, q_{1,4}^4, normal, \\ inflammatory, polyp, vascular \end{bmatrix} =$$

$$[0.51, 0.50, 0.39, 0.09, 0.02, 0.01, 0.14, 0.05, 0.41, 0.43, 0.29, 0.97, 0.03, 0.05, 0.01, 0.03, 0, 0, 0, 0].$$

Based on the defined fuzzy sets $\vartheta_{l,k}^{n,r}$ that represent linguistic terms, the initial state vector is fuzzified as follows: $A^0 =$ ["M", "M", "L", "VVL", "VVL", "VVL", "VVL", "VVL", "M", "M", "L", "VVH", "VVL", "VVL", "VVL", "VVL", "-", "-", "-", "-"]. The symbol "-" appearing in the last four values of A^0 corresponds to the initial output concept values, which for $t = 0$ have not yet been calculated. FCM begins the reasoning process, after A^0 is entered into the system, and iteratively evolves using

Eq. (1), until convergence. After 13 iterations, the FCM reaches a steady state, and converges; the respective convergence plot is illustrated in Fig. 2. The classification outcome is determined mainly by the last four values of A^{13} which correspond to the values of the output concepts:

$$A^{13} = \begin{bmatrix} 0.49, 0.48, 0.40, 0.05, 0.01, 0.03, 0.02, 0.06, 0.57, 0.54, 0.51, 0.98, 0.02, 0.06, 0.1, 0.04, \\ 0.48, 0.02, 0.94, 0.05 \end{bmatrix}.$$

Based on the defined fuzzy sets $\vartheta_{l,k}^{n,r}$ that represent linguistic terms, A^{13} takes the following form: $A^{13} =$ ["L", "L", "L", "VVL", "VVL", "VVL", "VVL", "VVL", "H", "H", "H", "VVH", "VVL", "VVL", "VVL", "VVL", "L", "VVL", "VVH", "VVL"]. Thus, comparing A^0 with A^{13}, the following observations can be made: (a) C_1 to C_4 corresponds to *"normal"* images with medoids ($q_{1,1}^1$ to $q_{1,1}^4$) and it has been redefined from ["M", "M", "L", "VVL"] to ["L", "L", "L", "VVL"]. This is because there is a negative influence among C_1-C_4 and $C_{18}-C_{20}$, as there exists a "VVL" similarity between the input image and the medoids of *"inflammatory"*, *"polyp"*, and *"vascular"*, respectively. Thus, an increase on C_1-C_4 evokes a decrease on C_{18}-C_{20}; as the influence of C_1-C_4 on C_{17} decreases, the risk of a lesion in the GI tract increases. Moreover, there is a positive influence among C_1-C_4 and C_{17}; this means that an increase in the input concepts occurs at the same time as an increase in C_{17}, when there is a "VVH" similarity of the input image and the corresponding medoids of "normal". Also it can be observed that: (b) $C_5 - C_8$ representing *"inflammatory (infl.)"* ($q_{1,2}^1 - q_{1,2}^4$) have the same values after 13 iterations ["VVL", "VVL", "VVL", "VVL"], given that there is a positive and "VVH" influence of $C_5 - C_8$ on C_{18}, and a negative and "VVL" causality of $C_5 - C_8$ on C_{17}, C_{19}, C_{20}; (c) $C_9 - C_{12}$, which are for *"polyp"* $\left(q_{1,3}^1 - q_{1,3}^4\right)$, had initially ["M", "M", "L", "VVH"] and are transformed to ["H", "H", "H" "VVH"]. Specifically, there is a positive and "VVH" interconnection between $C_9 - C_{12}$ and C_{19}, whereas the causality is negative and "VVL" among C_9-C_{12} with C_{17}, C_{18}, C_{20}; (d) $C_{13}-C_{16}$ that correspond to *"vascular (vasc.)"* ($q_{1,4}^1 - q_{1,4}^4$) have the same similarities as initially, *i.e.*, ["VVL", "VVL", "VVL", "VVL"]. The corresponding interconnections are positive and "VVH" among $C_{13} - C_{16}$ and C_{20}, whereas there is a negative and "VVL" influence of $C_{13} - C_{16}$ on $C_{17} - C_{19}$.

Furthermore, after 13 iterations, the output concepts $C_{17} - C_{20}$ corresponding to [*normal, inflammatory, polyp, vascular*] have values corresponding to the respective calculated similarities ["L", "VVL", **"VVH"**, "VVL"]. This means that there is a "VVH" risk of *"polyp"*, as there is "VVH" similarity between the features of the input image and the features of the image regions describing *"polyp"*, regarding their color and texture. Accordingly, the risk regarding *"normal"*, *"inflammatory"* and *"vascular"* is "L", "VVL", and "VVL", respectively.

3.3 Performance Evaluation of IF^3

The IF^3 framework was applied for lesion risk assessment in the GI tract. To estimate the effectiveness of IF^3, the proposed scheme was compared to other BoW-based models: Baseline [20], BoW-LBP [8], BoW-CLBP [8], Class-Specific BoW [11], Fuzzy Feature

Extraction BoW (FFE), and the explainable XFBoW model [11]. In addition, IF^3 was compared with the following CNN approaches: LB-FCN [6], WCNN [7]. The comparisons were performed in terms of the Area Under the Curve (AUC), Accuracy (Acc), Sensitivity (Sens), Specificity (Spec) [23]. The optimal FCM structure with which IF^3 framework achieved best results, was composed of 43 concept nodes. Table 1 includes the results of the comparisons of IF^3, regarding the KID Dataset 2 and the Kvasir. As it can be observed from Table 1, regarding the KID dataset 2, IF^3 outperforms almost all the compared methods, in terms of the AUC value, except LB-FCN. In addition, for Kvasir, IF^3 outperforms all the compared methods in terms of AUC. However, the advantage over LB-FCN is that IF^3 provides interpretable results, in a way understandable to human logic, while it is simple to implement. Also, IF^3 is characterized by a significantly lower execution time, compared to all the CNN-based methods for KID-Dataset 2, as it needs only 67 s, WCNN needs 18,660 s, and LB-FCN needs 7,200 s. To investigate the statistical significance of the performance of the proposed IF^3, in comparison to the rest methods, a two-tailed t-test [24] was used. In addition, the significance level used was set to 0.05. The calculated p-value evoked for all the examined cases, are smaller than 0.05. It can be concluded that there is a statistically significant difference, between the mean values of the examined methodologies. As a result, the null hypothesis, which describes the fact that the examined mean values are equal, is rejected.

Table 1. Performance evaluation of IF^3 using the datasets KID and Kvasir-v2.

Method/Metric	KID-Dataset 2				Kvasir-v2			
	AUC	Acc	Sens	Spec	AUC	Acc	Sens	Spec
XFBoW	0.84	0.81	0.48	0.89	0.93	0.89	0.86	0.92
WCNN	0.81	0.89	0.36	0.91	N/A	N/A	N/A	N/A
LB-FCN	0.93	0.88	0.92	0.76	N/A	N/A	N/A	N/A
Baseline BoW	0.80	0.76	0.45	0.88	0.90	0.85	0.81	0.89
Class-Specific	0.79	0.77	0.44	0.44	0.88	0.84	0.80	0.89
FFE	0.79	0.78	0.46	0.88	0.87	0.80	0.80	0.89
BoW-SIFT	0.72	0.74	0.31	0.87	0.83	0.78	0.61	0.90
BoW-CLBP	0.78	0.76	0.36	0.87	0.89	0.80	0.65	0.91
IF^3	0.86	0.82	0.63	0.93	0.95	0.90	0.87	0.92

4   Discussion and Conclusions

To address the challenge of automated detection of lesions in the GI tract, this paper introduces a novel AI framework based on FCMs. The proposed framework, IF^3, is generic, domain-independent and it can utilize features from different extractors. It constitutes a mechanism for automatic construction of an FCM based on similarities discovered between images of a training dataset. The proposed IF^3 was applied for lesion risk assessment in the GI tract, based on WCE images. To evaluate the performance of IF^3, experiments were conducted using real datasets with GI abnormalities, including polyps, which constitute precancerous lesions. The results of the experiments indicated that IF³ has mainly better or comparable performance, as compared with BoW-based and CNN-based approaches. However, a major advantage of IF³ is that it provides interpretable results, in a way compatible to human logic, based on its automated constructed graph structure. Furthermore, IF³ can be applied for any type of medical images, *e.g.*, ultrasound, computed tomography (CT) images, *etc.*

The proposed IF³ framework can be considered as an extension of XFBoW, with the aim of further improving the interpretability of the model based on a graph approach. IF³ gives the opportunity to the user to comprehend the reason why certain decisions or predictions have been made. This is because the proposed model has a transparent reasoning process that enables the user to monitor, in each iteration, how a decision is formed, until the model converges to a final inference. However, in the case of XFBoW the user can explain the result, without the ability to monitor the reasoning process. There are many benefits in using structured knowledge in the form of graphs to generate explanations, including their high understandability and accuracy [12]. Regarding FCMs, their simplicity and effectiveness successfully contributed to enhance the interpretability of the proposed approach, while providing outcomes in a way that is easily understandable from the perspective of users, without requiring any prior knowledge.

Future research includes the investigation of the effectiveness and time performance of the proposed framework, using other types of medical images and various extracted features, as well as one-dimensional signals, for applications such EEG analysis, and event prediction via telemonitoring of other biosignals.

Acknowledgment. We acknowledge support of this work by the project "Smart Tourist" (MIS 5047243) which is implemented under the Action "Reinforcement of the Research and Innovation Infrastructure", funded by the Operational Program "Competitiveness, Entrepreneurship and Innovation" (NSRF 2014-2020) and co-financed by Greece and the European Union (European Regional Development Fund).

References

1. Wang, S., Shen, L., Luo, H.: Application of linked color imaging in the diagnosis of early gastrointestinal neoplasms and precancerous lesions: a review. Ther. Adv. Gastroenterol. **14**, 17562848211025924 (2021)
2. Kim, D.H.: Other small bowel tumors. In: Chun, H.J., Seol, S.-Y., Choi, M.-G., Cho, J.Y. (eds.) Small Intestine Disease, pp. 243–248. Springer, Singapore (2022). https://doi.org/10.1007/978-981-16-7239-2_47
3. Dray, X., et al.: Artificial intelligence in small bowel capsule endoscopy-current status, challenges and future promise. J. Gastroenterol. Hepatol. **36**(1), 12–19 (2021)
4. Vasilakakis, M., Koulaouzidis, A., Yung, D.E., Plevris, J.N., Toth, E., Iakovidis, D.K.: Follow-up on: optimizing lesion detection in small bowel capsule endoscopy and beyond: from present problems to future solutions. Expert Rev. Gastroenterol. Hepatol. **13**(2), 129–141 (2019)
5. Painuli, D., Bhardwaj, S., et al.: Recent advancement in cancer diagnosis using machine learning and deep learning techniques: a comprehensive review. Comput. Biol. Med. 105580 (2022)
6. Diamantis, D.E., Iakovidis, D.K., Koulaouzidis, A.: Look-behind fully convolutional neural network for computer-aided endoscopy. Biomed. Signal Process. Control **49**, 192–201 (2019)

7. Iakovidis, D.K., Georgakopoulos, S.V., Vasilakakis, M., Koulaouzidis, A., Plagianakos, V.P.: Detecting and locating gastrointestinal anomalies using deep learning and iterative cluster unification. IEEE Trans. Med. Imaging **37**(10), 2196–2210 (2018)
8. Yuan, Y., Li, B., Meng, M.Q.-H.: Improved bag of feature for automatic polyp detection in wireless capsule endoscopy images. IEEE Trans. Autom. Sci. Eng. **13**(2), 529–535 (2015)
9. Angelov, P.P., Soares, E.A., Jiang, R., Arnold, N.I., Atkinson, P.M.: Explainable artificial intelligence: an analytical review. Wiley Interdiscip. Rev. Data Min. Knowl. Discov. **11**(5), e1424 (2021)
10. Prosperi, M., et al.: Causal inference and counterfactual prediction in machine learning for actionable healthcare. Nature Mach. Intell. **2**(7), 369–375 (2020)
11. Vasilakakis, M., Sovatzidi, G., Iakovidis, D.K.: Explainable classification of weakly anno-tated wireless capsule endoscopy images based on a fuzzy bag-of-colour features model and brain storm optimization. In: International Conference on Medical Image Computing and Computer-Assisted Intervention, pp. 488–498 (2021)
12. Tiddi, I., Schlobach, S.: Knowledge graphs as tools for explainable machine learning: a survey. Artif. Intell. **302**, 103627 (2022)
13. Kosko, B.: Fuzzy cognitive maps. Int. J. Man Mach. Stud. **24**(1), 65–75 (1986)
14. Felix, G., Nápoles, G., Falcon, R., Froelich, W., Vanhoof, K., Bello, R.: A review on methods and software for fuzzy cognitive maps. Artif. Intell. Rev. **52**(3), 1707–1737 (2017). https://doi.org/10.1007/s10462-017-9575-1
15. Mizumoto, M., Tanaka, K.: Fuzzy sets and their operations. Inf. Control **48**(1), 30–48 (1981)
16. Vasilakakis, M.D., Iakovidis, D.K., Spyrou, E., Koulaouzidis, A.: DINOSARC: color fea-tures based on selective aggregation of chromatic image components for wireless capsule endoscopy. Comput. Math. Meth. Med. **2018** (2018)
17. Koulaouzidis, A., et al.: KID Project: an internet-based digital video atlas of capsule endoscopy for research purposes. Endosc. Int. Open **5**(06), E477–E483 (2017)
18. Smedsrud, P.H., et al.: Kvasir-Capsule, a video capsule endoscopy dataset. Sci. Data **8**(1), 1–10 (2021)
19. Drake, J., Hamerly, G.: Accelerated k-means with adaptive distance bounds. In: 5th NIPS Workshop on Optimization for Machine Learning, vol. 8 (2012)
20. Vasilakakis, M., Iakovidis, D.K., Spyrou, E., Koulaouzidis, A.: Weakly-supervised lesion detection in video capsule endoscopy based on a bag-of-colour features model. In: Interna-tional Workshop on Computer-Assisted and Robotic Endoscopy, pp. 96–103 (2016)
21. Nápoles, G., Espinosa, M.L., Grau, I., Vanhoof, K.: FCM expert: software tool for scenario analysis and pattern classification based on fuzzy cognitive maps. Int. J. Artif. Intell. Tools **27**(07), 1860010 (2018)
22. Pelekis, N., Iakovidis, D.K., Kotsifakos, E.E., Kopanakis, I.: Fuzzy clustering of intuitionistic fuzzy data. Int. J. Bus. Intell. Data Min. **3**(1), 45–65 (2008)
23. Fawcett, T.: An introduction to ROC analysis. Pattern Recogn. Lett. **27**(8), 861–874 (2006)
24. Steel, R., Torrie, J., et al.: Principles and Procedures of Statistics. McGraw-Hill, New York (1960)

Lesion Characterization

A CAD System for Real-Time Characterization of Neoplasia in Barrett's Esophagus NBI Videos

Carolus H. J. Kusters[1](✉), Tim G. W. Boers[1], Jelmer B. Jukema[2],
Martijn R. Jong[2], Kiki N. Fockens[2], Albert J. de Groof[2],
Jacques J. Bergman[2], Fons van der Sommen[1], and Peter H. N. de With[1]

[1] Department of Electrical Engineering, Video Coding & Architectures,
Eindhoven University of Technology, Eindhoven, The Netherlands
c.h.j.kusters@tue.nl
[2] Department of Gastroenterology and Hepatology, Amsterdam University Medical
Centers, University of Amsterdam, Amsterdam, The Netherlands

Abstract. Barrett's Esophagus (BE) is a well-known precursor for Esophageal Adenocarcinoma (EAC). Endoscopic detection and diagnosis of early BE neoplasia is performed in two steps: primary detection of a suspected lesion in overview and a targeted and detailed inspection of the specific area using Narrow-Band Imaging (NBI). Despite the improved visualization of tissue by NBI and clinical classification systems, endoscopists have difficulties with correct characterization of the imagery. Computer-aided Diagnosis (CADx) may assist endoscopists in the classification of abnormalities in NBI imagery. We propose an endoscopy-driven pre-trained deep learning-based CADx, for the characterization of NBI imagery of BE. We evaluate the performance of the algorithm on images as well as on videos, for which we use several post-hoc and real-time video analysis methods. The proposed real-time methods outperform the post-hoc methods on average by 1.2% and 2.3% for accuracy and specificity, respectively. The obtained results show promising methods towards real-time endoscopic video analysis and identifies steps for further development.

Keywords: Barrett's Esophagus · NBI · Video analysis · Deep learning

1 Introduction

Esophageal adenocarcinoma (EAC) is a form of gastrointestinal cancer whose incidence has increased dramatically in the Western world [5]. Barrett's Esophagus (BE) is a well-known precursor of this form of cancer, posing a high risk for developing EAC. BE patients undergo regular endoscopic surveillance to enable

---

This work is facilitated by data/equipment from Olympus Corp., Tokyo, Japan.

© The Author(s), under exclusive license to Springer Nature Switzerland AG 2022
S. Ali et al. (Eds.): CaPTion 2022, LNCS 13581, pp. 89–98, 2022.
https://doi.org/10.1007/978-3-031-17979-2_9

treatment at an early stage with good prognosis [11,20]. However, the subtle endoscopic appearances [20] of neoplasia complicate its detection during surveillance
and sampling errors from random biopsies [6] make the surveillance protocol sub-
optimal. Endoscopic detection of neoplasia in BE is generally a process consisting
of two sequential steps. The first step is primary detection of a suspected lesion
using White-Light Endoscopy (WLE) in overview, followed by a targeted inspection of the mucosa, in order to find any visible abnormalities and to identify the
boundaries of the lesion. This targeted inspection is often performed by optical
chromoscopy techniques, such as Narrow-Band Imaging (NBI). The NBI imaging modality in combination with a magnified view enhances the visualization of
mucosal and vascular patterns. Several classification systems exploiting diverse
criteria have been proposed for NBI imagery of BE, but demonstrated to be sub-
optimal in terms of diagnostic accuracy and inter-observer agreement [1,9,16], so
that computer assistance may be attractive for improving diagnostic accuracy.

Artificial Intelligence (AI) with deep learning has emerged as a promising
technique to support medical professionals in gastrointestinal endoscopy [2–4,18,
23]. These supportive tools can be sub-divided into Computer-Aided Detection
(CADe) for finding neoplastic lesions and Computer-Aided Diagnosis (CADx) for
classification of these lesions into non-dysplastic or neoplastic. During the first
step of BE surveillance (detection of lesions), CADe tools may attractively benefit endoscopists. De Groof et al. [7,8] and Van der Putten et al. [13] designed a
highly effective CADe system for primary detection that recognized and localized
BE neoplasia in real time with high accuracy. This system is supplemented with a
CADx system to characterize tissue in the second step of BE surveillance. Such a
CADx system may improve the characterization performance, as well as dismiss-
ing false positive detections by the CADe system. The aim of this study is the
development of a deep learning-based CADx system for accurate characterization
of NBI imagery in BE after primary detection in WLE.

Several studies investigated the usage of deep learning-based CADx systems
in NBI imagery for colorectal polyp classification [17,22] or gastric lesion classification [10,21]. Van der Putten et al. [14] and Struyvenberg et al. [19] used a large
data set of NBI-zoom imagery to train a deep learning-based CADx system for
the classification of dysplasia in BE. In contrast to other work, endoscopic zoom
videos were employed, rather than still images, to improve the classification performance. Furthermore, the studies did not rely on ImageNet pre-training, but
built further upon endoscopy-driven pre-training introduced in their previous
work [15]. In other work, van der Putten et al. [12] identified challenges and
opportunities in endoscopic video analysis and addressed the first steps towards
clinical CAD application on endoscopic videos. However, this feasibility study
was based on WLE overview imagery rather than NBI imagery.

In this work, a significantly larger data set of both NBI imagery, consisting
of both still images and videos, is used for the classification of dysplasia in BE,
compared to the work of Van der Putten et al. [14] and Struyvenberg et al. [19].
However, in-line with both publications, endoscopy-driven pre-training is used in
our study. The proposed algorithm is developed using still images, after which
several video classification techniques are explored to classify BE dysplasia in

endoscopic videos. In contrast to previous work [12,14,19], which only employed post-hoc video analysis, in this work, we explore the first, necessary, steps towards an algorithm giving real-time feedback on endoscopic videos.

2 Methods

2.1 Data

1. Setting: Both retrospectively and prospectively collected data from 7 international centers are used in this study. All NBI imagery is obtained using HQ190 and EZ1500 gastroscopes (Olympus Corp., Japan) in combination with Excera-190 and X1-processors. Imagery of early BE neoplasia and non-dysplastic BE are contained in the dataset. Early BE neoplasia is defined as high-grade dysplasia or cancer in the corresponding endoscopic resection specimen, while non-dysplastic BE (NDBE) is histologically confirmed in all targeted biopsies.

2. NBI-Image Dataset: The algorithm is developed with 1,748 images of BE neoplasia (318 patients) and 1,762 images of NDBE (153 patients). Internal validation and hyper-parameter selection of the algorithm is performed with 45 images of BE neoplasia (23 patients) and 83 images of NDBE (31 patients). The external test set used for performance evaluation on unseen data, consists of 50 images of BE neoplasia (20 patients) and 111 images of NDBE (31 patients). In the splits made for testing and training/validation, separation on patient-basis is performed to avoid data leakage and intra-patient bias.

3. NBI-Video Dataset: The video validation set used for selection of algorithm video parameters, consists of 40 videos of BE neoplasia (19 patients) and 83 videos of NDBE (31 patients). The video test set consists of 50 videos of BE neoplasia (20 patients) and 111 videos of NDBE (31 patients). In both the validation and test video sets, videos of the same patients and areas of interests, as in their corresponding image set, are contained. For both the video validation and test set, the original version of the videos are available, as well as a version where non-informative frames (e.g. blur, bubbles, motion artifacts, illumination and contractions) are removed manually. The original videos of the validation set are in the range of 5–17 s, while the test set videos are in the range 7–10 s. The curated videos are in the range of 4–13 s and all videos take 4 s for the validation and test set, respectively.

4. Data Pre-processing: From raw endoscopic images, the central area is selected as the region of interest, such that the excessive black border is removed. The RGB images are resized to 256×256 pixels, before normalizing by channel-wise subtracting the mean and dividing by the standard deviation of the used training data. During training, the training set size is virtually increased by employing data augmentation techniques, in order to improve the generalization of the algorithm. The images are artificially corrupted with random Gaussian noise, followed by a random combination of horizontal and vertical flipping, rotation by $\theta \in \{0°, 90°, 180°, 270°\}$, contrast/saturation/brightness enhancements, gray-scale conversion, Gaussian blurring, random affine and sharpness transforms. In order to select hyper-parameters leading to a robust model, the internal validation set is sampled four times with each time a different random combination the first three previously mentioned augmentation techniques.

2.2   Network Architecture, Training and Evaluation

1. Network Architecture: The proposed network architecture has adopted the EfficientNet-B4 architecture and is further customized to improve suitability for the classification problem. The default classification head, consisting of 1,000 output neurons, is replaced by a customized version. This customized head consists of a flattened layer, followed by two fully-connected layers with 1,024 neurons and ReLU activation, finalized with a single-neuron fully-connected output layer using the Sigmoid activation. The EfficientNet-B4 framework was originally initialized with ImageNet pre-trained weights, after which a secondary pre-training is performed with an endoscopy-driven dataset. This dataset is known as GastroNet and described in work of Van der Putten *et al.* [15], consisting of $\approx 500,000$ gastro-intestinal endoscopic images. This pre-training stage enables the algorithm to learn features that are more representative for the target domain of endoscopic imagery, leading to an improved performance [15].

2. Training Procedure: The proposed network architecture is trained in two steps. First, the architecture is trained with all available training data for 75 epochs, or until convergence on the validation set. Second, after randomly initializing the weights of the custom classification head, the architecture is fine-tuned with only prospectively collected training images for 40 epochs, or until convergence on the validation set. The Adam optimizer with AMS-grad is used with learning rates of 10^{-4} and 10^{-5} for the first and second step, respectively, while $(\beta_1, \beta_2) = (0.9, 0.99)$ and a weight decay of 10^{-5} are used. During training, the batch size is 16 images and the loss of the algorithm is evaluated using the Binary Cross-Entropy loss function, with a loss penalty weight for positive samples of 3 and 5 for the first and second step of training, respectively. The proposed methods are implemented in Python using the PyTorch (Lightning) frameworks and experiments are executed on a GeForce RTX 3080 Ti.

3. Evaluation Metrics: The classification performance for videos and images is evaluated with the Accuracy (*Acc*), Sensitivity (*Sens*) and Specificity (*Spec*). The Accuracy is defined as the total percentage of correct predictions, Sensitivity and Specificity are the percentages of correctly predicted positive (BE Neoplasia) and negative (NDBE) images/videos, respectively. Finally, the Area under the Curve (*AUC*) is used for the receiver operating characteristic curve (*ROC*), summarizing the relationship between Sensitivity and Specificity.

2.3   Video Analysis Methods

1. Naive Approach: The frame-based performance of the algorithm on videos can be calculated, by comparing the frame prediction scores with a user-defined dysplasia threshold and obtaining a classification label. This single-frame performance is not a video-classification method, but will be reported to enable the comparison between the performance on high-quality still images and video frames, which are usually of lower quality. A multi-frame approach is adopted to incorporate temporal information embedded in the frame sequence and to obtain a full-video classification. Given the nature of the NBI inspection videos, where all frames have the same pathology, a video-classification score is obtained by simply taking the average of the frame-based prediction-score sequence. By comparing this average video-classification score with the dysplasia threshold, a classification label can be obtained.

2. Filter Approach: Not all video frames are informative and in contrast, the algorithm is mostly trained with high-quality still images and frames. The prediction scores for non-informative frames may induce noise in the prediction-score sequence. Therefore, a symmetrical filter $f_{M,\alpha}$ is proposed with the filter coefficients specified as: $f_{M,\alpha} = \frac{1}{M}[1, 2, \cdots, M/2-1, M/2, M/2-1, \cdots, 2, 1]^{\alpha}$, where M is the size of the filter and α denotes the weight factor for exponential decay. This non-causal filter $f_{M,\alpha}$ can be used to filter the prediction-score sequence $x[n]$ by employing convolution, giving $y[n] = \sum_m x[m] f_{M,\alpha}[n - m]$, where n is the frame index number and $y[n]$ is the filtered prediction-score sequence. Given these filtered prediction scores, the video-classification performance can be obtained by taking the average as explained in the naive approach.

3. Real-Time Approach: Real-time feedback is desired for clinical practice during video capturing, rather than giving post-hoc feedback. This requires real-time processing on the input video frame rate, which is not possible without latency, given the processing speed of the current algorithm and hardware. The proposed solution is to sub-sample the input video sequence. Thus, videos with a 50 fps are sub-sampled at 25 fps, while videos of 25 fps are analyzed at their input frame rate. We pursue to incorporate temporal information embedded in the frame sequence by adopting a multi-frame approach. We propose memorize the past L sub-sampled frame-prediction scores with a causal moving window of length L and applying a FIFO memory structure. Two options are considered to evaluate the classification label of the window. The first uses the window as a simple moving average (MA) filter, where the average prediction score of the window at each sub-sampled data point is calculated by $\mu_{\mathrm{MA}} = \frac{1}{L}\sum_m w[m]$, with L being the length of the window and $w[m]$ is the windowed frame prediction sequence. The window is extracted from the total frame prediction-score sequence $x[n]$ starting from position $n = i - L + 1$. The second option uses the window as a weighted moving average filter (WMA), where the latest data points are more important than older data points. In this case, the average prediction score of the window at each sub-sampled data point is calculated by multiplying each sample by its weight $(L, L - 1, \cdots, 2, 1)$, giving

$$\mu_{\mathrm{WMA}} = \frac{Lw[L - 1] + (L - 1)w[L - 2] + \cdots + 2w[1] + w[0]}{L + (L - 1) + \cdots + 2 + 1}. \tag{1}$$

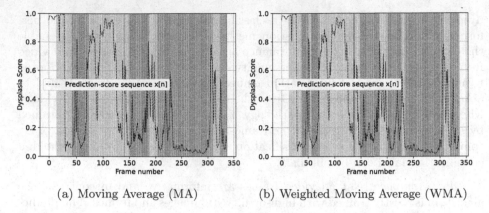

(a) Moving Average (MA)   (b) Weighted Moving Average (WMA)

Fig. 1. Two examples of prediction-score sequence $x[n]$ (dashed black line). The red and green backgrounds indicate the real-time classification label (neoplastic or non-dysplastic) of the causal moving window (L = 8, Dysplasia threshold=0.3). (Color figure online)

Table 1. Image- and frame-based classification performance.

Data	Curation	Validation set				Test set			
		Acc	$Sens$	$Spec$	AUC	Acc	$Sens$	$Spec$	AUC
Images	N/A	99.2%	100%	98.8%	0.998	93.8%	94.0%	93.7%	0.985
Frames	✗	93.8%	88.8%	96.1%	0.979	92.5%	91.5%	92.9%	0.975
Frames	✓	95.8%	90.5%	98.2%	0.991	93.2%	91.1%	94.1%	0.972

Subsequently, the average prediction scores calculated with μ_{MA} and μ_{WMA} are thresholded with the dysplasia threshold to obtain a classification label. The majority of the counts of individual classification labels (in this case neoplastic versus non-dysplastic) is decisive for the final video classification. Examples for prediction-score sequence $x[n]$ with corresponding real-time classification labels obtained with MA and WMA methods are illustrated in Fig. 1.

3   Experimental Results

1. Image- and Frame-Based Results: The classification performance results, using both still images and video frames of the original and curated videos, are presented in Table 1. To determine the classification label of a prediction, a dysplasia threshold of 0.3 is chosen, based on the performance on the image validation set. The image-based results show a decrease of 5.4%, 6.0%, 5.1% and 0.013 for *Acc*, *Sens*, *Spec* and *AUC*, respectively, when moving from the internal validation set (used for hyperparameter optimization) to the unseen test set. Furthermore, the results show that the frame-based performance is inferior to the image-based performance, for both the validation and test videos. A slight decrease of 1.3%, 2.5%, 0.8% and 0.01 for *Acc*, *Sens*, *Spec* and *AUC*, respectively, is observed for

Table 2. Video-based classification performance (Cur.=Curation).

Method	Cur.	Validation set				Test set			
		Acc	Sens	Spec	AUC	Acc	Sens	Spec	AUC
Naive Average	×	96.7%	95.0%	97.6%	0.998	91.3%	94.0%	90.1%	0.985
Filter + Average	×	96.7%	95.0%	97.6%	0.998	91.3%	94.0%	90.1%	0.985
Naive Average	✓	97.6%	95.0%	98.8%	0.997	93.8%	96.0%	92.8%	0.988
Filter + Average	✓	97.6%	95.0%	98.8%	0.997	94.4%	96.0%	93.7%	0.988

Table 3. Real-time video-based classification performance.

Method	Curation	Validation set			Test set		
		Acc	Sens	Spec	Acc	Sens	Spec
MA	×	96.7%	92.5%	98.8%	92.5%	94.0%	91.9%
WMA	×	96.7%	92.5%	98.8%	92.5%	94.0%	91.9%
MA	✓	97.6%	92.5%	100%	95.7%	96.0%	95.5%
WMA	✓	97.6%	92.5%	100%	95.0%	94.0%	95.5%

the original test video frames compared to the test images. Moreover, a modest increase of 0.7% and 1.2% for *Acc* and *Spec* is obtained, while a minor decrease of 0.4% and 0.003 for *Sens* and *AUC* is observed, when comparing the performance on the original test frames with the performance on the curated test frames.

2. Post-hoc Video-Based Results: The classification-performance results of the post-hoc naive average and filtered average approach, on original and curated videos, are presented in Table 2. A dysplasia threshold of 0.3, filter length of $M = 25$ and $\alpha = 0$ are chosen, based on the performance on the validation videos. The validation results show that filtering the prediction-score sequence before the averaging operation yields no extra performance, for both original and curated set. For original test videos, a performance gain is not achieved by the filtering operation, while for the curated test videos a minor increase is observed of 0.6% for *Acc* and 0.9% for *Spec*. Furthermore, the performances on curated videos are superior to the performances for original videos. The largest increase is observed for the combination of filtering and averaging on the test set videos of 3.1%, 2.0%, 3.6% and 0.003 for *Acc, Sens, Spec* and *AUC*, respectively.

3. Real-Time Video-Based Results The classification performance results of the real-time MA and WMA windowing approaches on both original and curated videos are presented in Table 3. A dysplasia threshold of 0.3 and a window length of $L = 8$ are chosen, based on the performance on the validation videos. The validation results show that there is no difference in classification performance between the MA and WMA windowing methods. The same holds for the original test videos, while for curated test videos the WMA method is slightly inferior to MA, with a minor decrease of 0.7% for *Acc* and 2.0% for *Sens*. The average processing time of the real-time approaches was 30 fps or 33 ms.

4   Discussion

The image-based performance of the proposed algorithm significantly outperforms the frame-based performance, especially in terms of sensitivity. The subjective image quality (e.g. blur, illumination, bubbles and motion artifacts) of the still images is ensured by the endoscopist when acquiring the image, while in a video this cannot be ensured for all frames. Furthermore, the algorithm is mostly trained with selected high-quality images and frames, so that it may not be sufficiently robust enough against quality variations. Therefore, removing the non-informative and low-quality frames in a video by manually curating and cutting videos, leads to a slightly increased performance compared to the original videos. However, manually removing the non-informative parts of a video is not feasible in clinical practice, requiring an informativeness assessment algorithm, as used in the study of Van der Putten *et al.* [12].

Post-hoc averaging of prediction scores for obtaining a single classification label for an entire video is superior in terms of classification performance, compared to employing a filtering operation prior to averaging. Sub-sampling the input video and using a causal moving window over the past L sub-sampled frames, enables real-time analysis on videos. Both real-time approaches outperform the post-hoc methods significantly, with an average increase of 1.2% and 2.3% for *Acc* and *Spec*, respectively. The MA method is preferred over the WMA method because the WMA method is more sensitive to outliers when using a small window size, as can be seen around frame numbers 50 and 230 in Fig. 1b.

Two limitations of the proposed real-time video classification methods can be identified. First, the methods can offer unstable output, as shown in Fig. 1, where the classification labels are fluctuating rapidly. Additionally, in order to define a stand-alone performance, the majority rule for obtaining the final class label of the entire video is arbitrarily chosen. This rule does not hold in clinical practice, because a delicate co-operation between endoscopist and algorithm exists in terms of the interpretation of multiple different class labels during video capturing. Second, the current algorithm operates on a large desktop, which is not feasible for implementation in endoscopy suites.

5   Conclusions

CADx systems for characterization of NBI imagery of BE neoplasia are clinically valuable to support endoscopists, who struggle with this task due to suboptimal clinical classification systems and the subtle endoscopic appearance of neoplasia. Current research towards deep learning-based CADx systems is mainly focused on post-hoc video analysis. In this work, we have explored the first steps towards an algorithm giving real-time feedback on endoscopic videos. We have investigated two methods for the real-time analysis of endoscopic video, which outperform post-hoc video analysis in terms of classification performance. However, several limitations of the proposed methods are identified. Therefore, future research should focus on the improvement of the output stability of the real-time analysis methods, to avoid rapidly fluctuating outputs and ensure a stable and reliable output. Furthermore, research should concentrate on smaller and efficient network architectures, which enable execution of the algorithm on edge devices or even implementation into existing endoscopy equipment. Lastly, incorporation of a frame-informativeness assessment or development of an endoscopic video quality-ensuring algorithm are virtually indispensable. In summary, this work offers the first promising steps towards automated real-time endoscopic video analysis and identifies steps for further development.

References

1. Baldaque-Silva, F., et al.: Endoscopic assessment and grading of Barrett's esophagus using magnification endoscopy and narrow band imaging: impact of structured learning and experience on the accuracy of the Amsterdam classification system. Scand. J. Gastroenterol. **48**(2), 160–167 (2013)
2. Byrne, M.F., et al.: Real-time differentiation of adenomatous and hyperplastic diminutive colorectal polyps during analysis of unaltered videos of standard colonoscopy using a deep learning model. Gut **68**(1), 94–100 (2019)
3. Chen, P.J., Lin, M.C., Lai, M.J., Lin, J.C., Lu, H.H.S., Tseng, V.S.: Accurate classification of diminutive colorectal polyps using computer-aided analysis. Gastroenterology **154**(3), 568–575 (2018)
4. Cho, B.J., et al.: Automated classification of gastric neoplasms in endoscopic images using a convolutional neural network. Endoscopy **51**(12), 1121–1129 (2019)
5. Dent, J.: Barrett's esophagus: a historical perspective, an update on core practicalities and predictions on future evolutions of management. J. Gastroenterol. Hepatol. **26**, 11–30 (2011)
6. Gordon, L.G., et al.: Cost-effectiveness of endoscopic surveillance of non-dysplastic Barrett's esophagus. Gastrointest. Endosc. **79**(2), 242–256 (2014)
7. de Groof, A.J., et al.: Deep-learning system detects neoplasia in patients with Barrett's esophagus with higher accuracy than endoscopists in a multistep training and validation study with benchmarking. Gastroenterology **158**(4), 915–929 (2020)
8. Groof, J., et al.: The Argos project: the development of a computer-aided detection system to improve detection of Barrett's neoplasia on white light endoscopy. United Eur. Gastroenterol. J. **7**(4), 538–547 (2019)
9. Kara, M.A., Ennahachi, M., Fockens, P., ten Kate, F.J., Bergman, J.J.: Detection and classification of the mucosal and vascular patterns (mucosal morphology) in Barrett's esophagus by using narrow band imaging. Gastrointest. Endosc. **64**(2), 155–166 (2006)
10. Lui, T.K., et al.: Feedback from artificial intelligence improved the learning of junior endoscopists on histology prediction of gastric lesions. Endosc. Int. Open **8**(02), E139–E146 (2020)
11. Pech, O., et al.: Long-term efficacy and safety of endoscopic resection for patients with mucosal adenocarcinoma of the esophagus. Gastroenterology **146**(3), 652–660 (2014)

12. van der Putten, J., et al.: First steps into endoscopic video analysis for Barrett's cancer detection: challenges and opportunities. In: Medical Imaging 2020: Computer-Aided Diagnosis, vol. 11314, p. 1131431. International Society for Optics and Photonics (2020)
13. van der Putten, J., et al.: Multi-stage domain-specific pretraining for improved detection and localization of Barrett's neoplasia: a comprehensive clinically validated study. Artif. Intell. Med. **107**, 101914 (2020)
14. van der Putten, J., et al.: Endoscopy-driven pretraining for classification of dysplasia in Barrett's esophagus with endoscopic narrow-band imaging zoom videos. Appl. Sci. **10**(10), 3407 (2020)
15. van der Putten, J., et al.: Pseudo-labeled bootstrapping and multi-stage transfer learning for the classification and localization of dysplasia in Barrett's esophagus. In: Suk, H.-I., Liu, M., Yan, P., Lian, C. (eds.) MLMI 2019. LNCS, vol. 11861, pp. 169–177. Springer, Cham (2019). https://doi.org/10.1007/978-3-030-32692-0_20
16. Sharma, P., et al.: Development and validation of a classification system to identify high-grade dysplasia and esophageal adenocarcinoma in Barrett's esophagus using narrow-band imaging. Gastroenterology **150**(3), 591–598 (2016)
17. Sierra, F., Gutiérrez, Y., Martínez, F.: An online deep convolutional polyp lesion prediction over narrow band imaging (NBI). In: 2020 42nd Annual International Conference of the IEEE Engineering in Medicine & Biology Society (EMBC), pp. 2412–2415. IEEE (2020)
18. Song, E.M., et al.: Endoscopic diagnosis and treatment planning for colorectal polyps using a deep-learning model. Sci. Rep. **10**(1), 1–10 (2020)
19. Struyvenberg, M.R., et al.: A computer-assisted algorithm for narrow-band imaging-based tissue characterization in Barrett's esophagus. Gastrointest. Endosc. **93**(1), 89–98 (2021)
20. Weusten, B., et al.: Endoscopic management of Barrett's esophagus: European society of gastrointestinal endoscopy (ESGE) position statement. Endoscopy **49**(02), 191–198 (2017)
21. Yan, T., Wong, P.K., Choi, I.C., Vong, C.M., Yu, H.H.: Intelligent diagnosis of gastric intestinal metaplasia based on convolutional neural network and limited number of endoscopic images. Comput. Biol. Med. **126**, 104026 (2020)
22. Zhang, R., et al.: Automatic detection and classification of colorectal polyps by transferring low-level CNN features from nonmedical domain. IEEE J. Biomed. Health Inform. **21**(1), 41–47 (2016)
23. Zheng, W., et al.: Tu1075 deep convolutional neural networks for recognition of atrophic gastritis and intestinal metaplasia based on endoscopy images. Gastrointest. Endosc. **91**(6), AB533–AB534 (2020)

Efficient Out-of-Distribution Detection of Melanoma with Wavelet-Based Normalizing Flows

M. M. Amaan Valiuddin$^{(\boxtimes)}$, Christiaan G. A. Viviers, Ruud J. G. van Sloun, Peter H. N. de With, and Fons van der Sommen

Eindhoven University of Technology, 5612 AZ Eindhoven, The Netherlands
m.m.a.valiuddin@tue.nl

Abstract. Melanoma is a serious form of skin cancer with high mortality rate at later stages. Fortunately, when detected early, the prognosis of melanoma is promising and malignant melanoma incidence rates are relatively low. As a result, datasets are heavily imbalanced which complicates training current state-of-the-art supervised classification AI models. We propose to use generative models to learn the benign data distribution and detect Out-of-Distribution (OOD) malignant images through density estimation. Normalizing Flows (NFs) are ideal candidates for OOD detection due to their ability to compute exact likelihoods. Nevertheless, their inductive biases towards apparent graphical features rather than semantic context hamper accurate OOD detection. In this work, we aim at using these biases with domain-level knowledge of melanoma, to improve likelihood-based OOD detection of malignant images. Our encouraging results demonstrate potential for OOD detection of melanoma using NFs. We achieve a 9% increase in Area Under Curve of the Receiver Operating Characteristics by using wavelet-based NFs. This model requires significantly less parameters for inference making it more applicable on edge devices. The proposed methodology can aid medical experts with diagnosis of skin-cancer patients and continuously increase survival rates. Furthermore, this research paves the way for other areas in oncology with similar data imbalance issues (Code available at: https://github.com/A-Vzer/WaveletFlowPytorch).

Keywords: Melanoma · Out-of-Distribution · Normalizing flows

1 Introduction

Melanoma, a form of skin cancer, develops in the melanocytes of the skin [3]. Symptoms can develop in the form of changing moles or growth of new pigmentation. Non-cancerous growth of the melanocytes is referred to as benign melanoma and is not harmful, while malignant melanoma is harmful. It is essential to recognize the symptoms of malignant melanoma as early as possible to classify its malignancy in order to avoid late diagnosis and ultimately an increased mortality rate [2]. To classify melanoma malignancy, experts consider indications of the

© The Author(s), under exclusive license to Springer Nature Switzerland AG 2022
S. Ali et al. (Eds.): CaPTion 2022, LNCS 13581, pp. 99–107, 2022.
https://doi.org/10.1007/978-3-031-17979-2_10

skin pigmentation such as asymmetrical shapes, irregular borders, uneven distribution of colors and large diameters (relative to benign melanoma) [1]. These clinical properties involve characteristics related to the texture and graphical details on the skin.

Since most cases of melanoma are benign, the number of malignant melanoma images are still relatively low. This data imbalance can negatively influence the predictions of machine learning (ML) models aiming to classify melanoma malignancy. Furthermore, most state-of-the-art supervised ML models are not calibrated, which poses the question on their validity for reliable skin-cancer detection [8]. Ideally, query images are assigned a calibrated confidence score, which can be interpreted as a probability of malignancy. Given these circumstances, a sensible option is to perform likelihood-based Out-of-Distribution (OOD) detection with the abundant benign data available.

Yielding tractable distributions, Normalizing Flows (NFs) serve as an excellent method for this application. NFs are a family of completely tractable generative models that learn exact likelihood distributions. However, OOD detection with NFs is notoriously difficult. This is caused by its inherent learning mechanisms that result in inductive biases towards graphical details, such as texture or color-pixel correlations rather than semantic context in images [12]. As such, OOD data is often assigned similar or higher likelihoods than the training data. In this paper, we show that with domain-level understanding of melanoma, we can improve NFs for OOD detection. Since the dominant features for indicating the malignancy of melanoma are described by their size and texture, we use wavelet-based NFs. We implement Wavelet Flow [15] for OOD detection of malignant melanoma and realize a 9% performance gain in Area Under Curve (AUC) of the Receiver Operating Characteristics (ROC). The number of parameters can significantly be reduced when applying Wavelet Flow for OOD detection, enabling implementation on smaller devices.

The proposed methodology presents the potential of NFs for aiding in reliable diagnosis of melanoma. Normalizing Flows for OOD detection and its inductive biases are discussed in Sect. 2. Thereafter, the approach and method is discussed in Sect. 3. The results are presented in Sect. 4 and concluded in Sect. 5.

2   Background

2.1   Normalizing Flows

Normalizing Flows are a sequence of bijective transformations, typically starting from a complex distribution, transforming into a Normal distribution. The log-likelihood $\log p(\mathbf{x})$ of a sample from the Normal distribution subject to an NF transformation $f_i : \mathbb{R} \mapsto \mathbb{R}$ is computed with

$$\log p(\mathbf{x}) = \log p_{\mathcal{N}}(\mathbf{z}_0) - \sum_{i=1}^{K} \log \left(\left| \det \frac{df_i}{d\mathbf{z}_{i-1}} \right| \right), \tag{1}$$

where the latent sample z_i is from the i-th transformation in the K-step NF and $p_\mathcal{N}$ the base Normal probability distribution. Due to the bijectivity of the transformations, Eq. (1) can be used to sample from $p_\mathcal{N}$ and construct a visual image with known probability. This transformation is referred to as the generative direction. An image can also be transformed in the normalizing direction (towards $p_\mathcal{N}$) to obtain a likelihood on the Normal density. Training in the normalizing direction is performed through Maximum Likelihood Estimation (MLE). Recently, many types of NFs have been proposed [4,7,9,11,13]. Better flows are generally more expressive, while having an computationally inexpensive Jacobian determinant. A widely used choice of NF are coupling flows, such as RealNVP and Glow [6,10]. The latter is used as a baseline for our experiments.

Out-of-Distribution Detection. The properties of NFs make them ideal candidates for OOD detection. Maximizing the likelihood of the data distribution $p(\mathbf{x})$ through a bijective transformation on $p_\mathcal{N}$ pushes away likelihoods of OOD data, when the density is normalized. Nevertheless, NFs assign similar likelihoods to train and (in-distribution) test data, indicating that flows do not overfit. This also indicates that not all OOD data receive low likelihoods. Ultimately, the assigned likelihoods are heavily influenced by the inductive biases of the model. Many NFs have inductive biases that limit their use for OOD detection applications [12].

Inductive Biases in Coupling Flows. Inductive biases of a generative model determine the training solution output and thus OOD detection performance. The input complexity plays an important role in OOD detection. Likelihood-based generative models assign lower likelihoods to more textured, rather than simpler images [14]. The widely accepted *affine coupling* NF is used in this research study. Kirichenko *et al.* [12] show that structural parts such as edges can be recognized in the latent space. This suggests that this type of flow focuses on visual appearance such as texture and color of the images, as opposed to the semantic content. Furthermore, the authors present coupling flow mechanics that cause NFs to fail at OOD detection. This is briefly discussed in order to keep the paper self-contained, but we encourage readers to refer to the original work. Given image $\mathbf{x}$, coupling flows mask it partly (x_m) and update it with parameters dependent on the non-masked part x_{res} as

$$x_m = (x_m + t(x_{res})) \cdot e^{s(x_{res})}, \tag{2}$$

where s and t are functions that output the scale and translation parameters, respectively. The log-Jacobian determinant in Eq. (1) for coupling flows is calculated as

$$\log\left(\left|\det\frac{df_i}{d\mathbf{z}_{i-1}}\right|\right) = -\sum_{i=1}^{D} s_i(x_{res}), \tag{3}$$

where i iterates over the image dimensionality D. Naturally, function s is encouraged to predict high values in Eq. (2) to maximize the log-likelihood in Eq. (1).

To compensate for this, function t must predict values that is an accurate approximation of $-x_m$. Therefore, the NF assigns high likelihood to images when the flow can accurately predict the masked part of the image. This can enable solutions that assign high likelihoods to any structured image, regardless of their semantic content. Two mechanisms are found to drive the accurate prediction of masked pixels and therefore assign higher likelihoods to OOD data. These are: learning local color-pixel correlations and information on masked pixels encoded in previous coupling layers, known as *coupling layer co-adaptation*. For the latter, different masking strategies such as cycle masking can be used to deprive the model from information in previous coupling-layer iterations [12]. Hence, we experiment with masking strategies to counteract coupling-layer co-adaptation. As an example with the opposite effect, checkerboard masking has been proposed [6]. Masking in this manner means that the predicted pixels are conditioned on its direct neighbouring pixels. Continuously, this encourages the NF to leverage local pixel correlations and further hinders semantically relevant OOD detection.

2.2 Wavelet Flow

Yu *et al.* [15] introduced the Wavelet Flow architecture (Fig. 1) for efficient high-resolution image generation. Instead of learning the image pixel likelihoods, the network models the conditional distribution with a coupling NF specified by

$$p(\mathbf{x}) = p(L_0) \prod_{i=0}^{N-1} p(D_i|L_i), \tag{4}$$

where D and L are the detail and low-frequency components of the Haar decomposition, respectively, and N represents the number of decompositions. During inference, an independent sample from $p(L_0)$ is up-scaled with the inverse Haar transform, using the predicted wavelet coefficients. To the best of our knowledge, this architecture has not been tested for OOD detection. Modeling the wavelet coefficients further guides the model to consider the graphical details of the image. As discussed in Sect. 1, melanoma can be distinguished by the texture of the skin. As a result, this inductive bias can improve OOD detection of melanoma. Furthermore, the high-frequency (detail) coefficients of the image enable easier distinguishment between highly textured malignant and less structured benign melanoma. This can facilitate better OOD detection, as NFs tend to assign higher likelihoods to smoother images.

3   Methods

As discussed in Sect. 2.1, the inductive biases of coupling NFs restrict their OOD detection capabilities. Given this information, we improve this by changing the data and model architecture. We test our approach on the ISIC dataset [5]. In this case, it can be beneficial that generative models assign higher likelihoods to

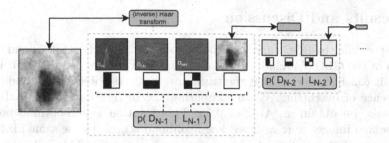

Fig. 1. Wavelet flow architecture. At each decomposition level, the likelihood of the high-frequency wavelet coefficients are learned conditioned on the low-frequency decomposition. Density $p(L_0)$ is modeled unconditionally.

less complex images, because benign melanoma are less textured and smaller in radius [1]. Initially, we downscale the RGB images to 128×128 pixels and train on the GLOW architecture naively, in a multi-scale setting, with default parameters $K = 32$ and $L = 3$. The AUC of the ROCs are used to evaluate the model performances. The color channels are heavily correlated and influence the likelihoods adversely, as discussed in Sect. 2.1. Therefore, we use grayscale images to hinder exploitation of local color-pixel correlations as well as to reduce training complexity. Thereafter, Wavelet Flow is employed. This shifts the optimization from the image pixels to their wavelet coefficients. This will further bias the model towards the graphical appearance of the images, since the tumor malignancy will be even more distinguishable by texture. Additionally, we experiment with different masking strategies (see Fig. 2). With Wavelet Flow, we obtain a likelihood, and thus an AUC score per decomposition scale. The individual likelihoods are averaged over all scales that contain sufficient information about the original content of the image. In this case, these are wavelet coefficients from 4×4 pixel dimensions up until the highest decomposition level. It might be beneficial to select only particular scales with good AUC values. However, this would constitute supervision, i.e. access to the malignant class, which is beyond the scope of this paper.

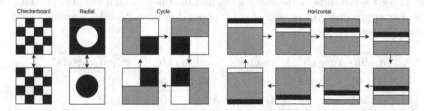

Fig. 2. Various masking strategies. The masks vary at each coupling flow step. The white area indicates the input of the s, t-network, which predicts parameters for the masked area in black. Grey areas are disregarded in the coupling process.

4   Results and Discussion

Table 1 presents the ROC curves for the various tested models. Likelihood distributions of the GLOW architecture trained on color images are shown in Fig. 3. Firstly, it can be observed that the train and test sets coincide well, indicating the absence of overfitting. When comparing the benign test to the malignant likelihoods, we obtain an AUC of 0.73. This solution is sub-optimal because many benign images were assigned low-likelihood scores. In the same likelihood range, most of the malignant images are present as well. This is because the model learns color-pixel correlations which can be used to leverage accurate predictions of the masked latent variables in the coupling layers. As a result, this leads to higher likelihoods assigned to OOD data.

When training on the wavelet coefficients with Wavelet Flow, there is substantial improvement on several decomposition scales (see Fig. 4). At all of the decomposition scales, besides the level seven (corresponding to the highest image resolution), we observe an improvement in test evaluation. We find the best AUC values from the 3rd up until the 6th decomposition scales. At these levels, the wavelet coefficients represent the most relevant frequency components of benign and malignant melanoma. As expected, the lowest decomposition scales contain almost no relevant information on the malignancy of melanoma and have very low AUC values. In Fig. 5, we average the likelihoods over the relevant decomposition scales. In a separate evaluation, we performed OOD detection using only the magnitude of the wavelet coefficients in which we observed acceptable AUC values on individual scales. However, in contrast with Wavelet Flow, averaging over the decomposition scales worked adversely. Furthermore, the different masking strategies did not improve performance.

Finally, some images are depicted of benign and malignant images of melanoma around various likelihoods (Fig. 6). We notice that at higher likelihoods, the malignant samples are more similar to that of the benign images. This indicates that the model is sufficiently learning relevant features, but is unable to classify early-phase malignant melanoma. As the likelihood values decrease, we observe more textured images. Specifically, larger pigmentation

Table 1. Test set results of the models trained on the ISIC dataset. For wavelet flow the number of parameters are that of the highest decomposition level as each level can be trained independently.

Architecture	K	L	channels	masking	ROC	# parameters*
GLOW	32	3	RGB	Affine	0.73	159M
GLOW	32	3	Gray	Affine	0.74	9.51M
GLOW	32	1	RGB	Affine	0.72	3.47M
GLOW	32	1	Gray	Affine	0.75	2.57M
Wavelet flow	32	1	Gray	All	0.78	2.50M
Wavelet flow	16	1	Gray	All	**0.78**	**1.25M**

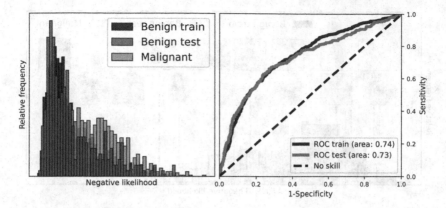

Fig. 3. Likelihood distribution and ROC curve of the trained GLOW architecture

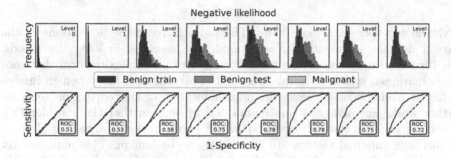

Fig. 4. Likelihood distributions per Haar wavelet decomposition level

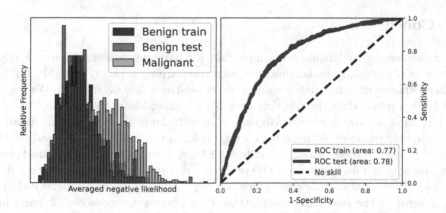

Fig. 5. The likelihood distribution and ROC curve of the trained Wavelet Flow architecture, averaged over the decomposition scales

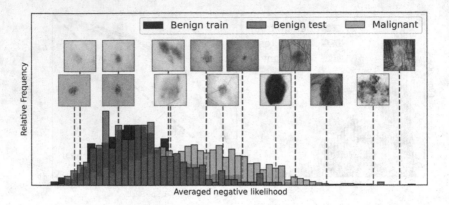

Fig. 6. Images of benign and malignant melanoma at various likelihoods. Note that lower likelihoods are either malignant or highly textured benign melanoma.

is visible together with more hairs. The hairs increase the activations in the wavelet domain, augmenting image complexity, resulting in lower likelihoods. The likelihood calculations can be corrected with a complexity term that considers hairiness, similar to Serrà *et al.* [14]. This will shift hairy benign images to higher likelihoods. We leave the implementation of this correction term for further work. For malignant melanoma, it can be seen that the consideration of texture goes beyond hairiness and size of pigmentation. Large skin pigmentations with minimal texture are more likely to be benign. This indicates yet again that the inductive biases of Wavelet Flow cause the model to sufficiently extract relevant information from the images.

5   Conclusion

Late diagnosis of melanoma poses high risks for patients with skin cancer. Early detection of malignant melanoma with machine learning is highly valuable, but is difficult due to data imbalance caused by its relatively low occurrence. We learn the benign image data distribution with Normalizing Flows to perform Out-of-Distribution (OOD) detection. We show that with knowledge on melanoma and the inductive biases of Normalizing Flows, we can improve likelihood-based OOD detection with wavelet-based Normalizing Flows. Furthermore, we demonstrate that memory requirements for OOD detection can significantly be reduced with Wavelet Flow, enabling the deployment on edge devices. We recommend including a term in the likelihood calculations that correct for presence of hairs in future work. The proposed methodology focuses solely on melanoma, however, we suggest that further research can facilitate exact likelihood-based OOD detection for other areas of oncology with large data imbalances to improve detection accuracy.

References

1. Melanoma. https://www.mayoclinic.org/
2. Melanoma survival rates. https://www.curemelanoma.org
3. What is melanoma skin cancer? (2022). https://www.cancer.org/cancer/melanoma-skin-cancer
4. Chen, R.T., Rubanova, Y., Bettencourt, J., Duvenaud, D.K.: Neural ordinary differential equations. Adv. Neural Inf. Process. Syst. **31**, 1–13 (2018)
5. Codella, N.C., et al.: Skin lesion analysis toward melanoma detection: a challenge at the 2017 international symposium on biomedical imaging (ISBI), hosted by the international skin imaging collaboration (ISIC). In: 2018 IEEE 15th International Symposium On Biomedical Imaging (ISBI 2018), pp. 168–172. IEEE (2018)
6. Dinh, L., Sohl-Dickstein, J., Bengio, S.: Density estimation using real NVP. arXiv preprint arXiv:1605.08803 (2016)
7. Durkan, C., Bekasov, A., Murray, I., Papamakarios, G.: Neural spline flows. Adv. Neural Inf. Process. Syst. **32**, 1–12 (2019)
8. Guo, C., Pleiss, G., Sun, Y., Weinberger, K.Q.: On calibration of modern neural networks. In: International Conference on Machine Learning, pp. 1321–1330. PMLR (2017)
9. Huang, C.W., Krueger, D., Lacoste, A., Courville, A.: Neural autoregressive flows. In: International Conference on Machine Learning, pp. 2078–2087. PMLR (2018)
10. Kingma, D.P., Dhariwal, P.: Glow: generative flow with invertible 1×1 convolutions. Adv. Neural Inf. Process. Syst. **31**, 1–10 (2018)
11. Kingma, D.P., Salimans, T., Jozefowicz, R., Chen, X., Sutskever, I., Welling, M.: Improved variational inference with inverse autoregressive flow. Adv. Neural Inf. Process. Syst. **29**, 1–9 (2016)
12. Kirichenko, P., Izmailov, P., Wilson, A.G.: Why normalizing flows fail to detect out-of-distribution data. Adv. Neural. Inf. Process. Syst. **33**, 20578–20589 (2020)
13. Rezende, D., Mohamed, S.: Variational inference with normalizing flows. In: International Conference on Machine Learning, pp. 1530–1538. PMLR (2015)
14. Serrà, J., Álvarez, D., Gómez, V., Slizovskaia, O., Núñez, J.F., Luque, J.: Input complexity and out-of-distribution detection with likelihood-based generative models. arXiv preprint arXiv:1909.11480 (2019)
15. Yu, J.J., Derpanis, K.G., Brubaker, M.A.: Wavelet flow: fast training of high resolution normalizing flows. Adv. Neural. Inf. Process. Syst. **33**, 6184–6196 (2020)

Robust Colorectal Polyp Characterization Using a Hybrid Bayesian Neural Network

Nikoo Dehghani[1]([✉]), Thom Scheeve[1], Quirine E. W. van der Zander[2,5],
Ayla Thijssen[2,5], Ramon-Michel Schreuder[3], Ad A. M. Masclee[2],
Erik J. Schoon[3,5], Fons van der Sommen[1,4], and Peter H. N. de With[1]

[1] Eindhoven University of Technology, 5612 AZ Eindhoven, The Netherlands
n.dehghani@tue.nl
[2] Maastricht University Medical Center+, 6229 HX Maastricht, The Netherlands
[3] Catharina Hospital, 5623 EJ Eindhoven, The Netherlands
[4] Eindhoven Artificial Intelligence Systems Institute,
5612 AZ Eindhoven, Netherlands
[5] GROW - School for Oncology and Reproduction, 6211 LK Maastricht, Netherlands

Abstract. Computer-Aided Diagnosis (CADx) systems can play a crucial role as a second opinion for endoscopists to improve the overall optical diagnostic performance of colonoscopies. While such supportive systems hold great potential, optimal clinical implementation is currently
impeded, since deep neural network-based systems often tend to overesti-
mate the confidence about their decisions. In other words, these systems
are poorly calibrated, and, hence, may assign high prediction scores to
samples associated with incorrect model predictions. For the optimal
clinical workflow integration and physician-AI collaboration, a reliable
CADx system should provide accurate and well-calibrated classification
confidence. An important application of these models is characterization of Colorectal polyps (CRPs), that are potential precursor lesions of
Colorectal cancer (CRC). An improved optical diagnosis of CRPs during the colonoscopy procedure is essential for an appropriate treatment
strategy. In this paper, we incorporate Bayesian variational inference and
investigate the performance of a hybrid Bayesian neural network-based
CADx system for the characterization of CRPs. Results of conducted
experiments demonstrate that this Bayesian variational inference-based
approach is capable of quantifying model uncertainty along with calibration confidence. This framework is able to obtain classification accuracy
comparable to the deterministic version of the network, while achieving
a 24.65% and 9.14% lower Expected Calibration Error (ECE) compared
to the uncalibrated and calibrated deterministic network using a post-
processing calibration technique, respectively.

Keywords: Colorectal polyp characterization · Bayesian inference ·
Model calibration · Classification uncertainty

© The Author(s), under exclusive license to Springer Nature Switzerland AG 2022
S. Ali et al. (Eds.): CaPTion 2022, LNCS 13581, pp. 108–117, 2022.
https://doi.org/10.1007/978-3-031-17979-2_11

1   Introduction

Colorectal cancer (CRC) ranks third in terms of most diagnosed cancer and appears as the second cause of cancer deaths in the world [1]. Colorectal polyps (CRPs) are precursor lesions of CRC and can be divided into two major categories, non-neoplastic and neoplastic. Non-neoplastic polyps, including Hyperplastic polyps (HP), are considered as benign polyps. In contrast, neoplastic polyps are consisting of the Adenomas (ADs) and Sessile Serrated Lesions (SSLs) and can harbor a malignant potential. It is possible to prevent CRC if these polyps are detected and removed at an early stage of the disease [2]. Colonoscopy is the most common procedure for screening and characterization of CRPs. Computer-aided diagnosis (CADx) systems can assist physicians with a more reliable diagnosis, by characterizing CRPs using optical methods.

With the advancement of deep neural networks, excellent results obtained by different CADx systems have been reported in literature for detection [3,4], segmentation [3,5,6], or classification [4,7–9] of CRPs. However, despite their recent success, these systems have not been extensively adopted in the clinical pilot studies so far. An important reason for the slow adoption of these systems is that neural networks are often over-confident in their decisions and fail to express the uncertainty over their predictions [10]. Thus, these systems may produce high class probabilities for incorrect predictions. These high-probability predictions can create harmful biases on physicians' decisions and become life-threatening in a clinical setting. Therefore, it is important that a model is capable of producing well-calibrated classification confidence along with its predictions.

Research on confidence calibration and the estimation of classification uncertainty, in the field of CRP characterization, has been limited. In [11,12], the authors investigated the roles of confidence calibration in CRP characterization via extra post-processing steps. As an alternative, alleviating the need for such additional training stages, Bayesian models have been widely adopted in different applications due to their ability to capture reliable uncertainty measures over the decision of the network during the training process, as evidenced by work of Krishnan *et al.* [13] for activity recognition. Bayesian neural networks (BNNs) [14,15] offer a probabilistic interpretation of deep learning models, by placing distributions over the model parameters and thereby learning from an ensemble of possible distributions of weights. Conversely, conventional Deep Neural Networks (DNNs) tend to disregard uncertainty around the model parameters by obtaining maximum likelihood estimates, which, in combination with most common loss functions, leads to overconfident decisions.

In this work, we propose a CADx system based on Bayesian variational inference [16] for characterization of CRPs. The system offers confidence calibration during the training procedure, in contrast to earlier studies on this topic [11,12], which require an extra post-processing step for the same purpose. Our results demonstrate that the proposed approach is not only competitive in terms of classification accuracy with respect to a Deterministic version of the model, but it is also able to provide reliable confidence measures. To the best of our knowledge, this is the first research study deploying a Bayesian variational inference framework for characterization of CRPs and expressing confidence-calibrated classification results.

2   Methodology

2.1   Dataset

The experiments conducted in this study are performed on data collected at the Catharina Hospital Eindhoven (CHE) and the Maastricht University Medical Center+ (MUMC+), in the Netherlands, and the Queen Alexandra Hospital (QA) in Portsmouth, United Kingdom. The dataset includes images with White-Light Endoscopy (WLE), Blue Light Imaging (BLI), and Linked Color Imaging (LCI)[1] modalities acquired from CHE and QA. Images collected at the MUMC+ have i-Scan modality in Modes 1, 2, and 3[2]. Several different polyp types are included, namely: HPs, ADs, SSLs and adenocarcinomas. The latter three polyp types are considered pre-malignant, and HPs are categorized as benign. In this study, experiments are carried out to classify CRPs into benign and (pre)malignant classes. To assess the classification performance of the proposed method, a total number of 2,287 images were used, including 1,836 pre-malignant polyps and 451 benign polyps. To evaluate the performance on unseen data, an independent test set is constructed, comprising 86 distinct polyps (258 images), of which 19 are benign and 67 pre-malignant. For each polyp, images from all three modalities are contained in the test set. The remainder of the data is split with 80/20% ratio for training and validation process, respectively, resulting into a training set of 316 benign and 1,308 pre-malignant images, while the validation set has 78 benign and 327 pre-malignant images. To prevent data leakage, all polyps from the same patient are kept together in one set (separation on patient basis).

2.2   Bayesian Neural Networks

Bayesian neural networks (BNNs) offer a probabilistic interpretation of deep learning models by learning a posterior distribution over the weights. As a result, the model will be robust to overfitting and is able to offer uncertainty estimates over the output probabilities.

In Bayesian statistics, network parameters are considered as one large random vector w, where the prior distribution of the weights is expressed as $p(w)$. If $X = \{x_1, ..., x_\beta\}$ denotes a set of training samples and $y = (y_1, ..., y_\beta)^T$ stands for the corresponding class labels, the posterior probability of the weights after observing the dataset is expressed as:

$$p(w|y, X) = \frac{p(y|w, X)p(w)}{\int p(y|w, X)p(w)dw}. \tag{1}$$

---

[1] EG-760 Colonoscope (Fujifilm® Corporation, Tokyo, Japan).
[2] EC38-i10F2 Colonoscope (PENTAX® Medical, Hoya Corp., Tokyo, Japan).

Classical assumptions on stochastic independence and modeling in deep learning, expresses the probability $p(y|w, X)$ as the product of the neural network outputs for all the training samples. Therefore, the integration over the very high-dimensional space of weights in the denominator of $p(y|w, X)$, makes the posterior generally intractable. Variational inference aims at approximating the posterior $p(w|y, X)$ by a distribution $(q_\Theta(w))$ that is most similar to the posterior distribution obtained by the model. This can be accomplished by Monte Carlo sampling of the posterior of model parameters and minimizing the Kullback-Leibler divergence *(KL-divergence)* between the variational distribution and the posterior: $D_{KL}(q_\Theta(w)||p(w|y, X))$.

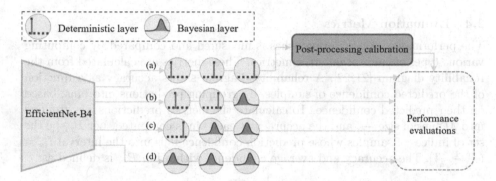

Fig. 1. Deterministic and hybrid Bayesian model architectures. On the left, the EfficientNet-B4 network is used as the base network and is followed by: (a) 3 fully-connected (FC) layers (addressed as the DNN); (b) 2 FC layers and 1 Bayesian layer; (c) 1 FC layer and 2 Bayesian layers; (d) 3 Bayesian layers (addressed as the BNN). The predictions made by the DNN are also passed to a post-processing calibration block. The outputs of all the networks are compared in the performance evaluation block.

2.3   Model Architecture

The block diagram of the employed framework is presented in Fig. 1. We use the EfficientNet-B4 architecture [17], pre-trained on ImageNet [18], as a base network and replace the classification layers with different sets of layers. As shown in Fig. 1 (a), the base architecture with 3 fully-connected (FC) layers serves as the Deterministic neural network (DNN).

The hybrid Bayesian models are achieved by gradually replacing each of the FC output layers with a Bayesian variational layer, which results in architectures (b) to (d) from Fig. 1. A Gaussian distribution is adopted to model the prior distribution of the weights and bias parameters in the Bayesian variational layers. During the training procedure, the aim is to minimize the KL-divergence by making multiple inference passes through the hybrid Bayesian networks. Inference passes are implemented using the Gradient Accumulation technique [19], to reduce the memory consumption. Flipout layers, as introduced by Wen *et al.* [20], serve as our Bayesian linear layers, due to their ability to decorrelate the gradients within a mini-batch as a result of implicitly sampling pseudo-independent weight perturbations for each data point.

In a first experiment, the calibration performance of the DNN is compared to the hybrid Bayesian model with 3 variational layers (BNN), using the reliability diagrams [21,22] and confidence measures that will be introduced later. A temperature scaling [10] method is also used to calibrate output results of the DNN (Calibrated DNN) to provide a better comparison with the BNN results. In another experiment, we will gradually adapt the DNN architecture towards a BNN by replacing its FC layers with variational layers, and evaluate the generalization and robustness property of the different degree of hybridization.

2.4   Evaluation Metrics

The performance of both approaches is measured and compared by computing various types of calibration error metrics. These metrics are calculated from the reliability diagram [21,22]. A reliability diagram shows accuracy as a function of the predicted confidence of samples, by grouping predictions into bins, based on their predicted confidence. To calculate the metrics predictions are grouped into M bins of size $\frac{1}{M}$, and the accuracy of each bin is computed. Let B_m be the set of indices of samples whose prediction confidence falls into the interval $I_m = (\frac{m-1}{M}, \frac{m}{M}]$. The accuracy and average confidence within bin B_m is defined as:

$$\text{acc}(\text{B}_m) = \frac{1}{|B_m|} \sum_{i \in B_m} 1(\hat{y}_i = y_i). \tag{2}$$

$$\text{conf}(\text{B}_m) = \frac{1}{|B_m|} \sum_{i \in B_m} (\hat{p}_i). \tag{3}$$

In the above equation, $\hat{y}_i$ and y_i are the predicted and true class labels, respectively, and $\hat{p}_i$ is the confidence for sample i. One important error metric is Expected Calibration Error (ECE), that is the weighted average of the calibration error across all bins. Moreover, the Maximum Calibration Error (MCE) determines the largest error across the bins. In line with the MCE, we also use the Average Calibration Error (ACE), that determines the average error across the non-empty bins (M^+). Finally, the Over-Confidence Error (OE) is specified as the weighted average of the errors across bins where confidence exceeds accuracy. These errors are covered by the following equations:

$$ECE = \sum_{m=1}^{M} \frac{B_m}{n} |\text{acc}(B_m) - \text{conf}(B_m)|. \tag{4}$$

$$MCE = \max_{m \in 1,\dots,M} \text{acc}(B_m) - \text{conf}(B_m)|. \tag{5}$$

$$ACE = \frac{1}{M^+} \sum_{m=1}^{M} \text{acc}(B_m) - \text{conf}(B_m)|. \tag{6}$$

$$OE = \sum_{m=1}^{M} \frac{B_m}{n} [\text{conf}(B_m) . \max(\text{conf}(B_m) - \text{acc}(B_m), 0)]. \tag{7}$$

3   Results

3.1   Experimental Setting

We evaluate our Bayesian network (BNN) in multiple experiments using the introduced CRP dataset and compare the obtained results with the uncalibrated and calibrated deterministic versions of the network (DNN). Input images for both networks are resized to 256×256 pixels, while compatibility of the dataset with ImageNet pre-trained networks is ensured by channel-wisely subtracting the mean and dividing by the standard deviation of ImageNet data. For an improved generalization, data augmentation is applied using the following transformations: horizontal and vertical flipping, rotation, Gaussian blurring, contrast/saturation/brightness enhancements, random affine, and perspective transforms. For the optimization, we use the Adam optimizer with a learning rate of 10^{-5}, $(\beta_1, \beta_2) = (0.9, 0.999)$. Due to the class imbalance, an independent batch generator is used to ensure that each of the classes is represented during training.

A mini-batch size of 16 (7 benign/9 pre-malignant) images is used and a the data is shuffled after each epoch. The training iteration ends when all benign images are seen once by the network. The experiments are implemented using the PyTorch framework and executed on a GeForce RTX 2080 Ti. To train the different hybrid BNN model variations, we perform multiple stochastic forward passes on the final (1–3) FC variational layers with Monte Carlo sampling on the weight posterior distributions. In our experiments, for a better generalization of the model, 10 forward passes provide reliable estimates. Subsequently, the predictive mean is obtained by averaging the confidence estimates from inference passes.

3.2   Calibration-performance Assessment

In order to verify the ability of the BNN to provide reliable confidence measures, we visualize and compare the calibration performance of the BNN with the DNN by using reliability diagrams. In these diagrams, the degree of miscalibration can be assessed by the gap between the plotted accuracy and the ideal diagonal. A high calibration performance can be achieved when the bin accuracy aligns closely with the ideal diagonal (expected accuracy). Figure 2 shows the reliability diagrams of the BNN, DNN, and the calibrated DNN with temperature scaling. The Green/Red bars indicate the Under/Over-Confidence, respectively. It can be observed that the BNN is better calibrated, as the achieved accuracies of the bins better approximate the expected accuracies (i.e. the bars align closer along the ideal diagonal). For the DNN and the calibrated DNN, the reliability diagrams show larger gaps between the achieved and the expected accuracies, especially for the higher confidence values.

Using the introduced calibration measures, a more quantitative comparison of the calibration performance of the three networks is achieved. Table 1 demonstrates a lower MCE of 0.2539 for the BNN compared to 0.2654 for the DNN,

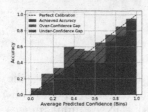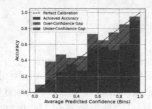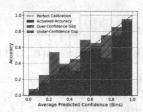

Fig. 2. Reliability diagrams for (left) the BNN, (center) the DNN, and (right) the calibrated DNN with temperature scaling.

Table 1. Calibration performance comparison for various experimented networks.

Network	ECE	MCE	ACE	OE
Bayesian network (BNN)	0.1699	0.2539	0.1296	0.0753
Deterministic network (DNN)	0.2255	0.2910	0.1721	0.0978
Calibrated deterministic network (Cal. DNN)	0.1870	0.2654	0.1388	0.0758

and 0.2910 for the calibrated DNN. In addition, the BNN network is able to achieve lower error rates using other calibration measures compared to the DNN as well as the calibrated DNN with the temperature scaling technique.

3.3 Model Performance Comparison

We evaluate our Bayesian and Deterministic models on the CRP dataset. In Table 2, a comparison of the obtained classification accuracy, Sensitivity, Specificity, area under curve (AUC), negative predictive value (NPV), and positive predictive value (PPV) on the test dataset are presented for each of the networks. The results show a very similar overall performance between the two networks, with most of the employed metrics exhibiting only a negligible difference.

3.4 Generalization and Robustness to Over-Fitting Assessment

In another experiment, we investigate the effect of increasing the Bayesian level of the deterministic network, by gradually replacing each of the FC layers of the Deterministic network by a Flipout variational layer, and obtain a network with increasingly more Bayesian layers (see Fig. 1 (b)–(d)), and finally obtain a network with 3 Bayesian FC layers. We have compared various versions of this

Table 2. Bayesian vs. Deterministic neural network performance assessment.

Network	Phase	Acc.	Sens.	Spec.	AUC	NPV	PPV
Bayesian network (BNN)	Test	84.10	90.55	61.40	0.89	64.81	89.22
Deterministic network (DNN)	Test	84.88	90.05	66.67	0.89	65.52	90.50

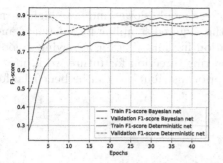 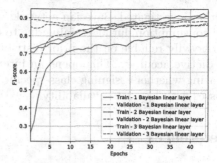

Fig. 3. Comparison of the F1-score during training and validation phase for (left) the Bayesian (BNN) and the Deterministic (DNN) networks, and (right) the different levels of hybridization for the Bayesian network.

hybrid BNN, in terms of F1-score of the training and validation phase and the result is available in Fig. 3 at the right. As demonstrated by the plots, the DNN shows a drop in validation F1-score as the training process advances, which shows that the model is over-fitting on the training data. On the other hand, the hybrid network with 3 Bayesian layers obtains a higher validation F1-score regarding its training F1-score and, therefore, offers a better generalized performance. Another important observation is that both networks with 1 and 2 Bayesian layers have a similar performance as the DNN, while experiencing the over-fitting problem to a lower degree. This possibly indicates the insufficiency of the Bayesian effect of the networks. It can be noticed that the network with 3 Bayesian layers (BNN) is capable of achieving comparable validation F1-score and expresses a better robustness towards over-fitting.

4   Discussion and Conclusion

Both an optimal clinical workflow integration and the physician-AI collaboration necessitate a reliable CADx system that is capable of capturing an accurate and well-calibrated classification confidence. In this regard, we incorporate Bayesian variational inference and investigate the performance of a hybrid Bayesian neural network architecture for the characterization of CRPs. The presented quantitative and qualitative results demonstrate that the BNN is capable of expressing reliable uncertainty measures and better calibrated classification confidence compared to a peer Deterministic network. Furthermore, the hybrid BNN approach is able to outperform a temperature-scaling calibrated DNN and provides lower calibration errors. Moreover, it alleviates the need for an additional calibration data set. A further hybridization experiment, based on replacing output layers with Bayesian variational layers, shows that the best performance is obtained by using 3 Bayesian layers. The better generalization property and being less prone to over-fitting, makes BNNs a suitable choice for small datasets. However, dealing with imbalanced classes can be an important challenge that should be further investigated. Bayesian networks are generally slower and have high memory consumption during training due to the required sampling for inference, and are heavily reliable on the prior distribution initialization for achieving a good predictive accuracy. This opens an interesting direction for future work, investigating whether assigning class-specific prior distributions can be beneficial for classes with less data availability.

References

1. Sung, H.,et al.: Global cancer statistics 2020: GLOBOCAN estimates of incidence and mortality worldwide for 36 cancers in 185 countries. CA Cancer J. Clinic. **71**(3), 209–249 (2021)
2. Shin, Y., Qadir, H.A., Aabakken, L., Bergsland, J., Balasingham, I.: Automatic colon polyp detection using region based deep CNN and post learning approaches. IEEE Access **6**, 40950–40962 (2018)
3. Meng, J., et al.: Automatic detection and segmentation of adenomatous colorectal polyps during colonoscopy using Mask R-CNN. Open Life Sci. **15**(1), 588–596 (2020)
4. Zhang, R., et al.: Automatic detection and classification of colorectal polyps by transferring low-level CNN features from nonmedical domain. IEEE J. Biomed. Health Inf. **21**(1), 41–47 (2016)
5. Wickstrøm, K., Kampffmeyer, M., Jenssen, R.: Uncertainty and interpretability in convolutional neural networks for semantic segmentation of colorectal polyps. Med. Image Anal. **60**, 101619 (2020)
6. Alam, S., Tomar, N.K., Thakur, A., Jha, D., Rauniyar, A.: Automatic polyp segmentation using u-net-resnet50. arXiv preprint arXiv:2012.15247 (2020)
7. Weigt, J., et al.: Performance of a new integrated computer-assisted system (CADe/CADx) for detection and characterization of colorectal neoplasia. Endoscopy. **54**(02), 180–184 (2022)
8. Usami, H., et al.: Colorectal polyp classification based on latent sharing features domain from multiple endoscopy images. Proc. Comput. Sci. **176**, 2507–2514 (2020)
9. Fonollà, R., et al.: A CNN CADx system for multimodal classification of colorectal polyps combining WL, BLI, and LCI modalities. Appl. Sci. **10**(15), 5040 (2020)
10. Guo, C., Pleiss, G., Sun, Y. and Weinberger, K.Q.: On calibration of modern neural networks. In: PLMR, International Conference on Machine Learning, pp. 1321–1330 (2017)
11. Kusters, K.C., et al.: Colorectal polyp classification using confidence-calibrated convolutional neural networks. In: SPIE, Medical Imaging 2022: Computer-Aided Diagnosis, vol. 12033, pp. 442–454(2022)
12. Carneiro, G., Pu, L.Z.C.T., Singh, R., Burt, A.: Deep learning uncertainty and confidence calibration for the five-class polyp classification from colonoscopy. Med. Image Anal. **62**, 101653 (2020)
13. Krishnan, R., Subedar, M. and Tickoo, O.: BAR: Bayesian activity recognition using variational inference. arXiv preprint arXiv:1811.03305 (2018)
14. Gal, Y., Ghahramani, Z.: Dropout as a Bayesian approximation: representing model uncertainty in deep learning. In: International Conference on Machine Learning, pp. 1050–1059. PMLR (2016)

15. Kendall, A., Gal, Y.: What uncertainties do we need in Bayesian deep learning for computer vision? Adv. Neural Inf. Process. Syst. **30**, 1–11 (2017)
16. Blundell, C., Cornebise, J., Kavukcuoglu, K., Wierstra, D.: Weight uncertainty in neural network. In: International Conference on Machine Learning, PMLR (2015)
17. Tan, M. and Le, Q.: Efficientnet: Rethinking model scaling for convolutional neural networks. In: International Conference on Machine Learning, pp. 6105–6114. PMLR (2019)
18. Deng, J., Dong, W., Socher, R., Li, L.-J., Li, K., Fei-Fei, L.: ImageNet: a large-scale hierarchical image database. In: Computer Vision and Pattern Recognition Conference, CVPR (2009)
19. Nazarovs, J., Mehta, R.R., Lokhande, V.S., Singh, V.: Graph reparameterizations for enabling 1000+ Monte Carlo iterations in Bayesian deep neural networks. In: Uncertainty in Artificial Intelligence, pp. 118–128. PMLR (2021)
20. Wen, Y., Vicol, P., Ba, J., Tran, D., Grosse, R.: Flipout: efficient pseudo-independent weight perturbations on mini-batches. arXiv preprint arXiv:1803.04386 (2018)
21. DeGroot, M.H., Fienberg, S.E.: The comparison and evaluation of forecasters. J. R. Statist. Soc. Ser. D (The Statist.) **32**(1–2), 12–22 (1983)
22. Niculescu-Mizil, A., Caruana, R.: Predicting good probabilities with supervised learning. In: 22nd International Conference on Machine Learning (2005)

Active Data Enrichment by Learning What to Annotate in Digital Pathology

George Batchkala[1]([✉]), Tapabrata Chakraborti[1], Mark McCole[2],
Fergus Gleeson[3], and Jens Rittscher[1]

[1] IBME/BDI, Department of Engineering Science, University of Oxford, Oxford, UK
{george.batchkala,tapabrata.chakraborty,jens.rittscher}@eng.ox.ac.uk
[2] Department of Cellular Pathology, Oxford University Hospitals NHS Trust,
Oxford, UK
mark.mccole@ouh.nhs.uk
[3] NCIMI/BDI, Department of Oncology, University of Oxford, Oxford, UK
fergus.gleeson@oncology.ox.ac.uk

Abstract. Our work aims to link pathology with radiology with the
goal to improve the early detection of lung cancer. Rather than utilising
a set of predefined radiomics features, we propose to learn a new set
of features from histology. Generating a comprehensive lung histology
report is the first vital step toward this goal. Deep learning has rev-
olutionised the computational assessment of digital pathology images.
Today, we have mature algorithms for assessing morphological features
at the cellular and tissue levels. In addition, there are promising efforts
that link morphological features with biologically relevant information.
While promising, these efforts mostly focus on narrow, well-defined questions. Developing a comprehensive report that is required in our setting
requires an annotation strategy that captures all clinically relevant patterns specified in the WHO guidelines. Here, we propose and compare
approaches aimed to balance the dataset and mitigate the biases in learning by automatically prioritising regions with clinical patterns underrepresented in the dataset. Our study demonstrates the opportunities active
data enrichment can provide and results in a new lung-cancer dataset
annotated to a degree that is not readily available in the public domain.

Keywords: Computational pathology · Histology annotation process ·
Unbalanced data · Image retrieval · Active and continual learning

1 Introduction

The ambition of finding new ways to link pathology and radiology in the context
of lung cancer motivates our goal of generating a comprehensive pathology report

GB is supported by FG and the EPSRC Center for Doctoral Training in Health Data
Science (EP/S02428X/1). TC is supported by Linacre College, Oxford. The work was
done as part of UKRI DART Lung Health Program.

Supplementary Information The online version contains supplementary material
available at https://doi.org/10.1007/978-3-031-17979-2_12.

© The Author(s), under exclusive license to Springer Nature Switzerland AG 2022
S. Ali et al. (Eds.): CaPTion 2022, LNCS 13581, pp. 118–127, 2022.
https://doi.org/10.1007/978-3-031-17979-2_12

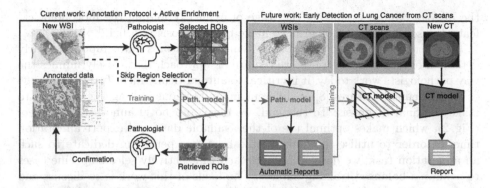

Fig. 1. Annotation protocol and active data enrichment (left) in the context of early detection of lung cancer from CT images (right). Trained pathology model will generate automatic histology reports for new WSIs. The generated reports will be used together with corresponding chest CT scans to learn a new set of radiology features in order to improve the early detection of lung cancer. Models in training are shown with sketch-style filling, while solid-fill represents trained models during the inference stage.

automatically. Given the current state of the art in computational pathology, this is an open problem.

Lung cancer accounts for more deaths than any other type of cancer [10]. The three main subtypes are Non-small Cell Lung Carcinoma (NSCC or NSCLC), Small Cell Carcinoma (SmCC), and Carcinoid Tumour. NSCLC accounts for more than 80% of all lung cancer cases [3,4] and is split into two main subtypes: lung adenocarcinoma (around 50% of all cases [7]) and lung squamous cell carcinoma. Based on existing clinical guidelines, CT is used to detect the presence of lung cancer. Tissue samples are then taken from suspicious regions to confirm the diagnosis. The general type, sub-type, and the underlying morphological characteristics of lung cancer determine clinical prognosis [8]. Hence it is vital to identify all subtypes, including those that occur less frequently, in a robust manner. The difficulty in making an accurate diagnosis lies in the inter- and intra-tumour heterogeneity [13]. The large inter-observer variability [9] is another factor that needs to be taken into account.

Recently a number of promising approaches for automatic subtyping of specific lung cancers and identifying specific lung-cancer morphologies have been published. With the help of extensive manual supervision in a form of region-based annotation, it is now possible to determine the predominant morphological pattern of lung adenocarcinoma [1,11]. Other works used WSI-level labels for weak supervision. Binary subtyping of NSCLC into adenocarcinoma and squamous cell carcinoma was performed in several works [2,5,6]. Yang *et al.* [12] extended it to six different types by adding samples from small cell lung carcinoma, pulmonary tuberculosis, organizing pneumonia, and normal lung tissue.

The drawback of all these methods is that they either identify morphologies of adenocarcinoma (the most prominent type of lung cancer) and do not take other types of lung cancer into account [1,11] or classify lung cancer types directly from the histology images omitting the stage of explicitly finding the morphological features used by the pathologists to make the diagnosis [2,5,6,12].

In order to support our goal of identifying new features that support the early detection of lung cancer on CT, we require an approach that closely mimics the way pathologists work today. It is critical to automatically identify a broad range of WHO-defined features [8] at different magnifications and aggregate them to make the final diagnosis. To this end, we develop a novel annotation protocol (Fig. 2), which makes optimal use of the available data and expert annotation time. In order to utilise the limited time human experts can dedicate to such an annotation task, we develop an approach that actively selects specific cases to achieve a balanced training dataset. The focus of this work is to discuss and analyse novel techniques to optimize the annotation process (Fig. 1, left).

2   Methodology

Our goal is to obtain region labels at different magnifications to support automated reporting of all clinically relevant subtypes of lung cancers. Here we take the WHO guidelines [8] as a reference. The labels should include the features and patterns used by the pathologists for making the diagnosis from a WSI.

2.1   Annotation Protocol

We propose a novel lung-cancer annotation protocol (Fig. 2) that consists of two main stages: (i) selection of relevant regions on digitised histology slides; and (ii) an annotation scheme that summarizes the information from the selected regions into a number of clinically-relevant patterns.

Stage 1: Region Selection. A typical digital slide viewer set-up is used to present the digitised slides. The pathologist is asked to mark enough diagnostically relevant regions on the slides in order to make the diagnosis by specifying regions of Interest (ROIs) (Fig. 2, left). Due to the small proportion of benign samples in our cohort, the pathologist also selects one or two regions of non-cancerous tissue on each of the slides to give explicit control of how a benign region can look and compensate for the lack of benign samples in the training data. Regions are mostly selected at two magnifications. Lower magnification allows to see *architectural patterns* (green ROIs), while higher magnification - *cytological features* (yellow ROIs).

Stage 2: Region Annotation. Aim of this step is to use the terms from the 2021 WHO Classification of Lung Tumours [8] to annotate each of the ROIs that have been selected at the previous stage. All relevant labels are shown on the right of Fig. 2. The "?" label is introduced to mitigate inter-observer variability. Only unequivocal cases are given definite labels.

2.2   Dataset Enrichment

Due to the high cost of expert annotation, it is vital to optimize the annotation process. For us, it means, minimizing the time spent by the pathologist in order

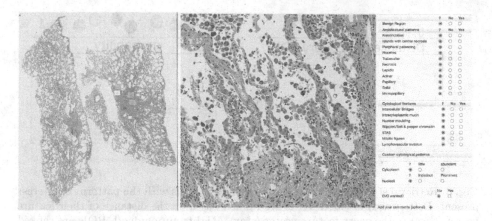

Fig. 2. Two stages of the annotation process. *Left*: pathologist is asked to choose a sufficient number of relevant regions of interest (ROIs) at different magnifications to support a diagnosis. *Right*: annotation view for one of the ROIs. (Color figure online)

to achieve the quality of the data enough for efficient model training. When training a model to recognize multiple classes at once, it is crucial to create a balanced dataset in which all classes are well represented.

A naive sequential annotation of the available data would naturally result in an extremely unbalanced dataset in which patterns of rare disease subtypes would be underrepresented. Trying to get a sufficient number of regions with under-represented patterns in this naive sequential way would result in sub-optimal use of limited expert annotation time. To avoid this, we propose two approaches to increase the proportion of regions with underrepresented patterns and help the models learn features distinguishing these patterns by making use of known image retrieval techniques. The approaches are illustrated on Fig. 1 (left): ranking regions pre-selected by the pathologist (solid arrows) and automatically selecting regions from non-annotated WSIs in a sliding-window sweep (dashed line, planned future work). We have tried supervised and unsupervised methods for the former approach (described in Sect. 3 to avoid repetition). As a result, we are now in a position to capture a unique reference dataset for lung histology.

In this work, we consider the recognition and enrichment independently for different patterns. However, in reality, groups of patterns can coexist at certain magnifications. This might cause a problem: *if* (1) pattern A and pattern B coexist at the same magnification and are usually seen together; (2) pattern A is easier to spot than pattern B for our retrieval method, *then* the retrieval method can learn to predict the presence of pattern B when pattern A is present and disregard any features specific to pattern B. Currently, we need more samples for all the chosen morphological features. Our work is a proof-of-concept study showing that it is possible to enrich for certain morphological patterns using the proposed methods if needed. The label distribution for the pre-selected regions is provided in the Supplementary Material.

As our data collection is ongoing, we do not know how many annotated data points can be generated. Hence we propose a metric to measure the retrieval

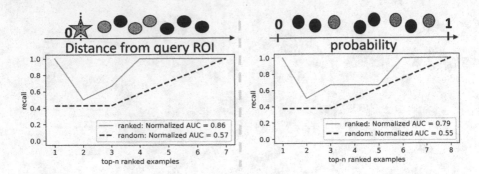

Fig. 3. Retrieval strategies. Textured dots represent ROIs with the patterns of interest present. **Left: unsupervised**. ROIs are ranked based on the distance of their feature vector (dots) to the query feature vector (star). **Right: supervised**. ROIs are ranked by their probabilities of having the pattern of interest present.

performance for a variable number of examples: **Ranking Curve AUC**. Our metric is similar to ROC AUC used for measuring classification performance. For each n the ranking curve shows the proportion of regions with the pattern of interest in top-n ranked samples from the total number of regions with the pattern of interest that are possible to pick up in n samples $(min(n, t))$, where t is the total number of relevant examples. We chose this denominator since we can not possibly find more relevant examples in any n samples. For the AUC calculation, discrete data points of the ranking curve are connected with straight lines. The normalized version of the proposed metric has an upper bound of 1 when all relevant samples are ranked higher than irrelevant ones. See the formal definition and the derivation of properties in Supplementary Material. Figure 3 shows ranking curves for toy examples.

3   Results

We compare unsupervised, supervised, and active-learning retrieval strategies for ranking regions pre-selected by the pathologist.

3.1   Unsupervised Data Enrichment

To study the effectiveness of the enrichment methods, a pathologist annotated 2 batches with 20 and 17 WSIs that resulted in 145 and 120 selected ROIs, respectively. The batches were scanned with different scanners at 20× magnification. Keratinization is the most underrepresented in the first batch, with only 1 of the ROIs marked to have it by our pathologist. Having only one image of a particular class suggests unsupervised image-retrieval since we can not train a good model from one example. Furthermore, this region does not have any other patterns from our annotation list. Hence, we enrich the keratinization class as follows. We use the region as a *query region*. It is passed through a feature extractor to create a *query vector*. Not-annotated *target regions* pre-selected by the pathologist are processed by the same feature extractor. Matches with the smallest cosine similarity, a distance metric commonly used for image retrieval, are presented to the pathologist for confirmation.

Table 1. Retrieval performance of different feature extractors. The proportion of regions with keratinization in the second batch is $15/120 = 1/8$ meaning that we expect $20 * 1/8 = 2.5$ examples with keratinization in any random sample of 20 examples.

Feature extractor	Pre-training	Success rate	Ranking AUC
Modified ResNet-50 [6]	ImageNet	3/20	0.62
ResNet-18 w/o last layer	ImageNet	4/20	0.62
ResNet-18 w/o last layer	TCGA patches at 2.5× [5]	7/20	0.73
ResNet-18 w/o last layer	TCGA patches at 10× [5]	**8/20**	**0.79**
Random	NA	2.5/20	0.51

15 out of 120 ROIs in the second batch contain keratinization pattern. We simulate the situation of prioritising which of the 120 patterns we should annotate to maximize the number of images with keratinization in the top-20 ranked samples since it is the number of ROIs that the pathologist reported being able to annotate in one annotation session. Since we do not require extra annotation, we try different feature extractors: a modified version of an ImageNet pre-trained ResNet-18 without the classification layer, an ImageNet pre-trained modified ResNet-50 [6], and two ResNet-18 models pre-trained using self-supervised learning on patches extracted at 2.5× and 10× magnifications from TCGA-lung by Li *et al.* [5] (results in Table 1). The TCGA-lung pre-trained networks perform better than the ImageNet pre-trained ones. However, 20 is still an arbitrary number of ROIs to consider, and this choice can seriously affect the performance evaluation. Hence, we use the Ranking Curve AUC proposed in Sect. 2.2. The curves for two feature extractors are shown in Fig. 4 while the AUC values are presented in Table 1. ResNet-18 feature extractor pre-trained on patches of TCGA-lung extracted at 10× magnification [5] showed best retrieval results. For any 40 regions chosen at random, we expect to have $40/120 \approx 0.33$ or 5 out of 15 regions with keratinization pattern. However, the top-40 ranked regions have 10 out of 15 regions doubling the proportion compared to random choosing ($10/15 \approx 0.66$). Only with a third of annotated regions from the second batch, we could get the labels for two-thirds of the regions with keratinization pattern.

3.2 Supervised Active Data Enrichment

We now explore how using supervised enrichment methods can be used to prioritise the regions with acinar pattern present. We evaluate how including these regions improves the classifier performance on a previously unseen test set and mitigates the biases in learning by reducing the class imbalance.

124   G. Batchkala et al.

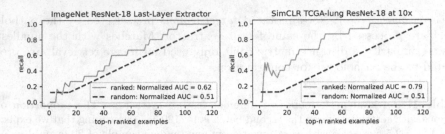

Fig. 4. Solid orange line: ranking curve (as described in Sect. 2.2). Dashed blue line: expected cumulative proportion if selecting samples at random. ImageNet trained extractor (left) shows worse results than TCGA-lung pre-trained extractor (right). (Color figure online)

We split the dataset into 4 subsets: Train, Validation, Pool, and Test with 20/86, 14/59, 10/60, and 10/60 examples containing the acinar pattern, respectively (see Supplementary Material for detailed data distribution). Pool imitates the regions next in line for annotation. Train, Validation, and Test sets serve their usual roles. Train and Validation sets come from the first batch of images, while Pool and Test sets come from the second batch. The batches were scanned with different scanners. This shows how in real life we can have training and evaluation data coming from different distributions.

To obtain a baseline, we train a single classification layer on top of a frozen ResNet-18 feature extractor pre-trained on patches from TCGA-lung at 10× magnification [5] on the Train set using Cross-Entropy loss with 3 possible labels for the acinar pattern: "yes", "no", "not sure". When calculating Cross-Entropy loss, we use the same weights for all classes since the label distribution changes in different batches, and we want to be able to predict all of them well in the end. We report unweighted accuracy, ROC AUC weighted by the number of support samples with each label, as well as Ranking Curve AUC described in Sect. 2.2, precision, and recall for the "yes" label. We save the weights of the models which showed the best ROC AUC on the validation set. Weighted ROC AUC is chosen because it is sensitive to class imbalance, gives a good understanding of the model performance, and is a popular choice in the literature [2,5,6,11,12].

			Accuracy	ROC AUC	Rank AUC	Pre (yes)	Re (yes)
86		Train	0.65	0.82	0.752	0.333	0.2
86	10×10	Train + 10 Rand	0.75 ± 0.028	0.83 ± 0.026	0.82 ± 0.036	0.5 ± 0.191	0.39 ± 0.158
86	10	Train + 10 Rank	0.8	0.864	0.874	0.667	0.6
86	20×10	Train + 20 Rand	0.75 ± 0.04	0.83 ± 0.018	0.83 ± 0.042	0.59 ± 0.164	0.46 ± 0.143
86	20	Train + 20 Rank	0.783	0.904	0.923	1.0	0.2
86	30×10	Train + 30 Rand	0.76 ± 0.031	0.84 ± 0.017	0.85 ± 0.011	0.55 ± 0.097	0.57 ± 0.1
86	30	Train + 30 Rank	0.817	0.924	0.954	0.833	0.5
86	60	Train + Pool	0.833	0.865	0.894	0.636	0.7

Fig. 5. Experiments were conducted by training the model on different training sets. For each $N \in \{10, 20, 30\}$, adding N ranked examples results in better models then adding N random examples. For random selection, 10 sets of N examples were taken from the Pool set with mean ± one standard deviation reported for each metric.

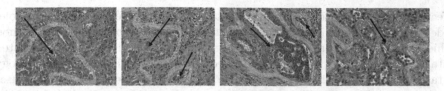

Fig. 6. Regions containing acinar pattern from the top-10 ranked pool set examples returned by our method. Solid arrows point at areas confirmed and delineated by the pathologist to contain acinar patterns, thus validating the results.

Having the baseline model, we vary the training data by including different portions of the Pool set into it. Given that the "yes" label is underrepresented in the first batch and our particular interest in learning to predict it better, we propose to rank the samples by sorting them in decreasing order of predicted probabilities of the "yes" label. We assess how including 10, 20, and 30 ranked examples from the Pool affects the performance metrics. For comparison, we repeat the experiments with 10, 20, and 30 randomly-chosen examples from the Pool. To account for randomness when choosing a subset of examples, we repeat the experiments 10 times for each subset size. Finally, we include all 60 Pool set examples to get the largest-training data baseline (results in Fig. 5).

We observe that including more $(0 \rightarrow 10 \rightarrow 20 \rightarrow 30)$ ranked or non-ranked samples into the training data increases both the performance of the classifier (weighted ROC AUC) and the ability of the classifier to rank the regions with acinar pattern higher than the ones without it (Ranking Curve AUC). Furthermore, adding 10 ranked examples improves ROC AUC, Ranking Curve AUC, and Precision more than adding 10, 20, or even 30 random examples. Finally, including all 60 Pool examples improves accuracy and recall but results in lower weighted ROC AUC, Ranking AUC, and Precision. We believe that this happens because we optimize the parameters of the network using a non-weighted Cross-Entropy loss which optimizes the weights better for more prevalent classes. The proportion of the "yes" class in Train + Pool is 0.205, while it is 0.25. 0.245, 0.241, for Train + 10, +20, +30 Ranked images respectively (see data distribution in Supplementary Material). This change in proportion results in the improved unweighted accuracy but removes the precise focus from the "yes" class, which hurts ROC AUC (sensitive to class imbalance), Ranking AUC and Precision, which are all measured for the "yes" class.

Figure 6 shows examples from the top-10 ranked samples from the Pool set. The samples were ranked using predicted probabilities of the acinar pattern presence. These four regions were later confirmed and delineated by the pathologist to contain the acinar pattern, thus adding further validation to our method.

4   Conclusion

We present a new comprehensive annotation protocol for lung histopathology. In order to increase the value of expert annotations, we propose a simple yet effective method for prioritizing the annotation of regions extracted from whole-slide images that are likely to contain underrepresented patterns. The method achieves this by utilising a region-retrieval model created using the annotated data. The proposed method is evaluated on a new in-house lung-pathology dataset. We conclude that even with little supervision, we can enrich the dataset for a pattern of interest. This method is now being used to actively balance the distribution of labels in the annotated portion of our dataset. This, in turn, will result in better classification and retrieval models. We plan to use better retrieval models to bypass the region selection stage (Fig. 1, left, dashed line). Finally, we will use the classification models for automatic histology report generation. The generated reports will be used to learn a new set of radiology features from histology in order to improve the early detection of lung cancer (Fig. 1, right).

References

1. Alsubaie, N., Shaban, M., Snead, D., Khurram, A., Rajpoot, N.: A multi-resolution deep learning framework for lung adenocarcinoma growth pattern classification. In: Nixon, M., Mahmoodi, S., Zwiggelaar, R. (eds.) MIUA 2018. Communications in Computer and Information Science, vol. 894, pp. 3–11. Springer, Cham (2018). https://doi.org/10.1007/978-3-319-95921-4_1
2. Coudray, N., et al.: Classification and mutation prediction from non–small cell lung cancer histopathology images using deep learning. Nat. Med. **24**(10), 1559–1567 (2018). https://doi.org/10.1038/s41591-018-0177-5
3. Davidson, M.R., Gazdar, A.F., Clarke, B.E.: The pivotal role of pathology in the management of lung cancer. J. Thorac. Dis. **5**(Suppl 5), S463-478 (2013). https://doi.org/10.3978/j.issn.2072-1439.2013.08.43
4. Huang, T., et al.: Distinguishing lung adenocarcinoma from lung squamous cell carcinoma by two hypomethylated and three hypermethylated genes: a meta-analysis. PLoS ONE **11**(2), e0149088 (2016). https://doi.org/10.1371/journal.pone.0149088
5. Li, B., Li, Y., Eliceiri, K.W.: Dual-stream multiple instance learning network for whole slide image classification with self-supervised contrastive learning. In: Proceedings of the IEEE/CVF Conference on Computer Vision and Pattern Recognition (CVPR), pp. 14318–14328 (2021). https://doi.org/10.1109/CVPR46437.2021.01409
6. Lu, M.Y., Williamson, D.F., Chen, T.Y., Chen, R.J., Barbieri, M., Mahmood, F.: Data-efficient and weakly supervised computational pathology on whole-slide images. Nat. Biomed. Eng. **5**(6), 555–570 (2021). https://doi.org/10.1038/s41551-020-00682-w
7. Meza, R., Meernik, C., Jeon, J., Cote, M.L.: Lung cancer incidence trends by gender, race and histology in the United States, 1973–2010. PLoS ONE **10**(3), 1–14 (2015). https://doi.org/10.1371/journal.pone.0121323
8. Nicholson, A.G., et al.: The 2021 WHO classification of lung tumors: impact of advances since 2015. J. Thorac. Oncol. Off. Publ. Int. Assoc. Study Lung Cancer **17**(3), 362–387 (2022). https://doi.org/10.1016/j.jtho.2021.11.003
9. Stang, A., et al.: Diagnostic agreement in the histopathological evaluation of lung cancer tissue in a population-based case-control study. Lung Cancer **52**(1), 29–36 (2006). https://doi.org/10.1016/j.lungcan.2005.11.012

10. Torre, L.A., Siegel, R.L., Jemal, A.: Lung cancer statistics. In: Ahmad, A., Gadgeel, S. (eds.) Lung Cancer and Personalized Medicine. AEMB, vol. 893, pp. 1–19. Springer, Cham (2016). https://doi.org/10.1007/978-3-319-24223-1_1
11. Wei, J.W., et al.: Pathologist-level classification of histologic patterns on resected lung adenocarcinoma slides with deep neural networks. Sci. Rep. **9**(1), 3358 (2019). https://doi.org/10.1038/s41598-019-40041-7
12. Yang, H., et al.: Deep learning-based six-type classifier for lung cancer and mimics from histopathological whole slide images: a retrospective study. BMC Med. **19**(1), 80 (2021). https://doi.org/10.1186/s12916-021-01953-2
13. Zhao, B., et al.: Reproducibility of radiomics for deciphering tumor phenotype with imaging. Sci. Rep. **6**(1), 23428 (2016). https://doi.org/10.1038/srep23428

Segmentation, Registration, and Image-Guided Intervention

Comparing Training Strategies Using Multi-Assessor Segmentation Labels for Barrett's Neoplasia Detection

Tim G. W. Boers[1]([✉]), Carolus H. J. Kusters[1], Kiki N. Fockens[2],
Jelmer B. Jukema[2], Martijn R. Jong[2], Jeroen de Groof[2], Jacques J. Bergman[2],
Fons van der Sommen[1], and Peter H. N. de With[1]

[1] Eindhoven University of Technology, Eindhoven, The Netherlands
t.boers@tue.nl
[2] Amsterdam UMC, Amsterdam, The Netherlands

Abstract. In medical imaging, segmentation ground truths generally suffer from large inter-observer variability. When multiple observers are used, simple fusion techniques are typically employed to combine multiple delineations into one consensus ground truth. However, in this process, potentially valuable information is discarded and it is yet unknown what strategy leads to optimal segmentation results. In this work, we compare several ground-truth types to train a U-net and apply it to the clinical use case of Barrett's neoplasia detection. To this end, we have invited 14 international Barrett's experts to delineate 2,851 neoplastic images derived from 812 patients into a higher- and lower-likelihood neoplasia areas. Five different ground-truths techniques along with four different training losses are compared with each other using the Area-under-the-curve (AUC) value for Barrett's neoplasia detection. The value used to generate this curve is the maximum pixel value in the raw segmentation map, and the histologically proven ground truth of the image. The experiments show that random sampling of the four neoplastic areas together with a compound loss Binary Cross-entropy and DICE yields the highest value of 94.12%, while fusion-based ground truth clearly performs lower. The results show that researchers should incorporate measures for uncertainty in their design of networks.

Keywords: Interobserver variance · Segmentation · Neural networks

1 Introduction

A long-standing challenge in medical imaging is the utilization of data sets that lack a gold-standard ground truth for multiple observers. In medical segmentation tasks, often ground truth are based on subjective expert opinions and this leads to high inter-observer variability in a different range of tasks [3–5, 8, 10].

This work is facilitated by data/equipment from Olympus Corp., Tokyo, Japan.

© The Author(s), under exclusive license to Springer Nature Switzerland AG 2022
S. Ali et al. (Eds.): CaPTion 2022, LNCS 13581, pp. 131–138, 2022.
https://doi.org/10.1007/978-3-031-17979-2_13

One application for which this phenomenon is clearly present is endoscopic neoplasia detection in Barrett's esophagus (BE). The edges of the neoplastic area are specifically hard to determine, and oftentimes not even distinguishable in the white-light endoscopy image, leading to a considerable inter-observer disagreement among experts. This inter-observer variability can be broadly noticed in the MICCAI 2015 Endoscopic Vision Challenge (data described in [9]), where experts only agreed $79.35 \pm 7.6\%$ on relatively easy-to-detect lesions. As a consequence, the usage of non-optimal segmentation labels for training a segmentation model can lead to a degraded performance, and can therefore deteriorate the detection rate of Barrett's lesions [11]

In order to address inter-observer variability in medical image segmentation, supervised learning approaches for deterministic models are typically trained using a ground truth that is generated by common fusion techniques (e.g. majority voting). Most studies focusing on segmentation of Barrett's Neoplasia also use these fusion techniques. De Groof *et al.* [2] define their ground truth for segmentation training as the intersection area from the annotations of 4 experts for training their segmentation algorithm. Ebigbo *et al.* [1] use the intersection area of 5 experts as the ground truth for their algorithm. Van der Putten *et al.* [7] use multiple data sets, which were annotated by 2–4 experts, where the area with one or more expert delineations served as ground truth for training and validation.

Although fusion of ground-truths from multiple observers might intuitively seem better than single-observer annotations for training, it has limitations for true clinical practice. Due to the subtle gradation of changes that occur in the progression of dysplasia in BE, a wide range of morphological patterns of atypical features are observed, which are not necessarily neoplastic, but *are* part of the transition process in the nature of the tissue [6]. In practice, this causes a significant degree of intra-observer and inter-observer variability in the diagnosis of dysplasia, particularly for subtle cases of neoplasia [6]. Although fusion techniques are commonly proposed in literature [2] to address variability, they yield binary segmentation masks and thereby treat these areas as either fully neoplastic or fully non-dysplastic Barrett's esophagus (NDBE). This approach fails to capture the subtle nuance of neoplasia gradation needed for the detection of particularly subtle variants of neoplasia.

Currently, little is known about how inter-observer variability, lesion subtlety and commonly used fusion methods affect the performance of deterministic image-based detection algorithms. Research on uncertainty estimation is ongoing based on multi-observer labeling exploiting Bayesian modeling in neural networks. However, since these networks are relatively slow due to repeated sampling, they are less suitable for real-time operation. In order to facilitate efficient real-time deterministic inferencing, we postulate that a supervised learning approach needs to jointly reflect the disagreement among experts together with a gradation of neoplasia, in order to improve the quality and clinical value of Barrett's lesion detection.

In this work, we propose and investigate several fusion techniques of consensus ground truths for training segmentation models, in order to optimize the detection accuracy of such models. For this, the studied use case is on Barrett's neoplasia detection, but it is pursued that our findings will be applicable to other fields. We hypothesize that when a deterministic neural network is trained using ground truths reflecting both the subtle gradation of neoplasia and observer variability, the network will improve its detection abilities, hopefully in quantity and quality.

2 Materials and Methods

2.1 Data Set

A dedicated data set is assembled of endoscopic images for training, validating and testing, with histologically proven ground truth. For each data set, images originating from a single patient are only assigned to one set. Each data set is carefully assembled by clinicians, in order to be representative for the various tumor characteristics described by the Paris classification. De-identification is performed in line with the General Data Protection Regulation (EU) 2016/679.

The training set consists of 2,651 neoplastic images (708 patients) and 7,595 NDBE images (1,095 patients), the validation set contains 100 neoplastic images (54 patients) and 100 NDBE images (36 patients). Finally, the test set contains 100 neoplastic images (50 patients) and 300 NDBE images (125 patients).

Each of the single neoplastic images are delineated by two experts on Barrett's neoplasia, which both delineate two areas within the image. The first area is referred to as the Higher-Likelihood (HL), which contains the area that the expert considers definitely to be neoplasia. The second area is referred to as the Lower-Likelihood (LL), which looks atypical from normal NDBE tissue, but for which the expert is not certain that it contains neoplasia. In total, 14 international experts have contributed to the total data set, who helped to include various annotation styles. For the HL neoplasia delineation, we have used a minimum of 30% DICE score overlap between experts, since this ensures that there is at least a minimal consensus about the neoplastic area among the experts. If the DICE score is less than 30%, then a third expert is invited to annotate the sample as a tiebreaker. The two out of three delineations with the highest overlap are then used as the ground truth.

2.2 Segmentation Ground-truth Assembly

Per image we have obtained four segmentations, two HL and two LL. Five different forms of ground truth are compared for training the network model, which are further detailed below. Also, an example of each form of ground truth is illustrated in Fig. 1.

Softspot: The Softspot is defined as the union of all four delineations. Therefore, in the segmentation ground truth, this mask will label all neoplastic and atypical tissue as being neoplastic.

Sweetspot: The Sweetspot is defined as the union of the two HL delineations. This segmentation ground truth is labeling all pixels neoplastic that are considered neoplastic by either of the two observers.

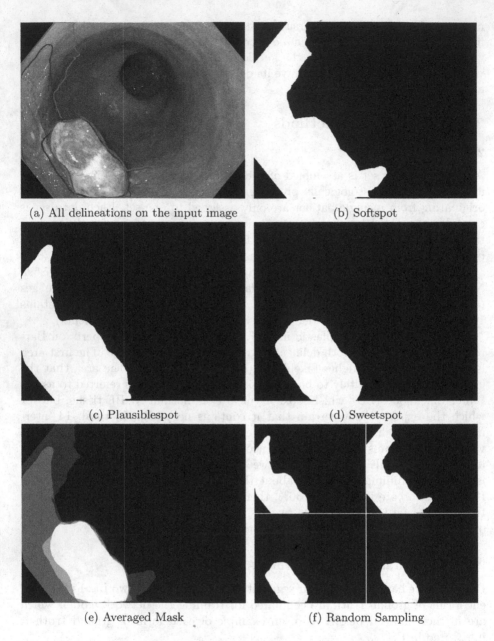

Fig. 1. Example case of neoplasia delineation and various forms of ground-truth strategies used to train our model to detect Barrett's neoplasia. (a) Input image, with
the thick dark red and blue lines, representing the HL per observer, and the thinner
lines, in red and blue, representing the LL. The red and blue colors are also matched
with their respective observer. (b)–(e) Proposed forms of the generated segmentation
ground truths. (f) All four segmentation masks, from which one is randomly chosen per
iteration. (Color figure online)

Plausiblespot: The Plausiblespot is the union of the Sweetspot with the intersection of the two LL delineations. This is an intermediate form of the Sweetspot and the Softspot. Here, the segmentation includes the area that either of the two observers labeled as neoplastic and the area that both observers consider to look atypical.

Averaged Maps: For the Averaged maps method, all segmentation maps are pooled into a single map with the average values. This will yield a ground truth that develops gradually with non-binary values r for indicating neoplasia, with $r \in \{0.0, 0.25, 0.5, 0.75, 1.0\}$ as ground truth.

Random Sampling: In the Random sampling approach, each segmentation map is randomly sampled and provided to the model as ground truth during the training process. This is a form of data augmentation at the output of the neural network. Therefore, the neural network has to learn a generalized form of all used segmentations by itself.

2.3   Network Architecture

A joint ImageNet-pretrained ResNet-18 encoder, combined with a U-net decoder model is chosen as CNN architecture. The ResNet structure is adopted for the encoder because of its efficiency, while the U-net decoder stands out in medical segmentation. Max-pooling is used to downsample the encoder, while nearest-neighbor interpolation is employed to upsample feature maps in the decoder. Standard Rectified Linear Units are used for all activation functions, except for the activation function after the final convolutional layer. This output is supplied to a Sigmoid function to generate continuous output values in the unity range. Furthermore, Batch Normalization is used after each convolutional layer.

2.4   Training Details

Networks and Losses: For our experiments, the U-net architecture is trained using 18 different configurations. This architecture is trained with five different ground truths, as described in Sect. 2.2, and four different loss functions. These loss functions are: mean-squared error (MSE), binary cross-entropy (BCE), DICE and a compound loss BCE+DICE. However, for the Averaged-maps ground truth we only use two loss functions, since a DICE loss is only suitable for binary ground-truth labels. The experiment is repeated 5 times with different random seeds, in order to create more generic results.

Learning Rate: The algorithm is trained using Adam and AMS-grad with a weight decay of 10^{-4}. A cyclic cosine learning-rate scheduler is used to control the learning rate. Additionally, batch normalization is employed to regularize the network.

Data Augmentation: Images and segmentation masks have been randomly rotated with $\theta \in \{0°, 90°, 180°, 270°\}$ and randomly flipped along the $x-$ and

y-axis with probability $p = 0.5$. Additionally, random permutations have been made to the contrast, brightness and saturation of the images.

Data Balancing: Since the training set is class-imbalanced, the training images are randomly sampled such that each class is represented 50% on average during each iteration. The model is trained for 16,100 iterations with a batch size of 64 using float16 precision.

3   Experiments and Results

3.1   Metrics

During training, the accuracy of the model is measured on the validation set and the model with the highest score is chosen as the final model. After training, we evaluate the detection performance of the model by calculating the Receiver-operating-characteristic Area-Under-the-Curve (ROC-AUC) value on the unseen test set. In order to yield a single value for a multi-digit segmentation mask, we have simplified the output by taking the maximum value from the segmentation mask. This value is then compared to the histologically proven classification label, in order to generate the ROC-AUC curve.

3.2   Results

A meta-analysis of the training data set show that the international experts only agree $59.6 \pm 25.4\%$ DICE for HL neoplasia area and $71.6 \pm 20.4\%$ DICE for LL neoplasia area. On average, the HL segmentations are $11.9 \pm 9.7\%$ of the number of pixels, and the LL are on average $25.8 \pm 16.2\%$ of the image pixels.

Each model has been retrained five times with a different initialization. The mean results of these experiments, along with the standard deviation between brackets, are reported in Table 1. A plot of an ROC-AUC curve from one training cycle is shown in Fig. 2, in order to highlight the spread between training configurations. The figure shows that five ground-truth types have a considerable spread in performance, where several proposed forms provide high performance, albeit in specific intervals.

Table 1. Averaged AUC values for five ground-truth options, obtained after 5 full training cycles. The values between the brackets denote the standard deviations.

Ground-truth type	MSE	BCE	DICE	BCE+DICE
Softspot	90.23 (1.06)	90.07 (1.96)	91.33 (1.03)	90.47 (1.82)
Plausiblespot	88.23 (1.35)	88.48 (1.54)	89.86 (1.98)	88.54 (1.09)
Sweetspot	90.27 (1.71)	88.64 (2.85)	91.31 (1.62)	93.28 (1.19)
Random sampling	93.83 (1.00)	92.52 (1.62)	92.76 (0.75)	**94.12 (0.89)**
Averaged maps	92.23 (1.73)	92.77 (1.60)	N/A	N/A

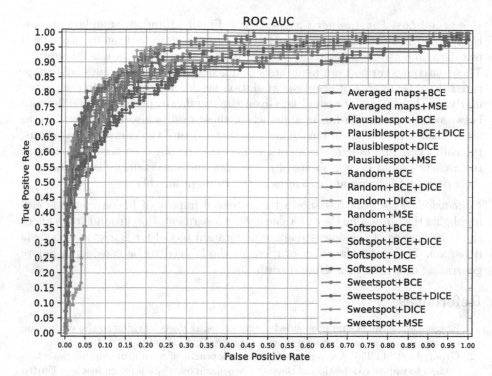

Fig. 2. ROC-AUC curve of the 18 different experimental configurations. This graph shows the test results of a single cycle for simplicity.

4   Discussion and Conclusions

We have analyzed the impact of multi-observer fusion and provisioning of Higher-likelihood (HL) and Lower-likelihood (LL) segmentation ground truths on the detection performance using modern neural networks for Barrett's neoplasia.

Spread in Observer Ground Truth: The meta-analysis shows that experts in the field of Barrett's neoplasia tend to have a relatively low agreement on the HL area of neoplasia with only 59.6% DICE score. This implies that the total data set contains more difficult lesions for segmentation, and supports the claim from the introduction that a notable level of disagreement exists between experts on the delineation of neoplasia in endoscopic images.

Non-fusion Methods are Better: Moreover, we have also observed that all non-fusion-based segmentation ground truths lead to a better detection performance compared to fusion-based generated ground truths (sweetspot, plausiblespot and softspot). This implies that taking some form of gradation into the segmentation ground truth, either via random sampling or averaging of the segmentation maps, facilitates the network model to better learn generic features representing the complex nuance of neoplasia seen in endoscopic BE imaging.

Compound Loss Function for Complex Data: Finally, Random Sampling in combination with a compound BCE and DICE loss generalizes best to the unseen test data, and given the relatively low variance, also provide reproducible results. These outcomes support our hypothesis that general fusion techniques fail to capture the subtle nuance of neoplasia gradation needed for the detection of particularly subtle variants of neoplasia. Given the significant performance difference between fusion and non-fusion approaches, these results also likely translate to other fields where a single binary label cannot capture the complex gradations, typically occurring in image contents. In general, the obtained results show that researchers should be well aware of the nature and uncertainty distributions of their data and incorporate measures in their design for this.

Conclusion: This work has shown the potential impact of fusion methods and highlights that fusion methods for generating a segmentation ground truth affect the performance of the downstream segmentation model for Barrett's neoplasia detection. Therefore, the effect of these methods need to be considered before generating of a consensus ground truth.

References

1. Ebigbo, A., et al.: Computer-aided diagnosis using deep learning in the evaluation of early oesophageal adenocarcinoma. Gut **68**(7), 1143–1145 (2019)
2. Groof, J., et al.: The Argos project: the development of a computer-aided detection system to improve detection of Barrett's neoplasia on white light endoscopy. United Eur. Gastroenterol. J. **7**(4), 538–547 (2019)
3. Joskowicz, L., Cohen, D., Caplan, N., Sosna, J.: Inter-observer variability of manual contour delineation of structures in CT. Eur. Radiol. **29**(3), 1391–1399 (2019)
4. Lazarus, E., Mainiero, M.B., Schepps, B., Koelliker, S.L., Livingston, L.S.: BI-RADS lexicon for us and mammography: interobserver variability and positive predictive value. Radiology **239**(2), 385–391 (2006)
5. Menze, B.H., et al.: The multimodal brain tumor image segmentation benchmark (brats). IEEE Trans. Med. Imaging **34**(10), 1993–2024 (2014)
6. Odze, R.: Diagnosis and grading of dysplasia in Barrett's oesophagus. J. Clin. Pathol. **59**(10), 1029–1038 (2006)
7. van der Putten, J., et al.: Pseudo-labeled bootstrapping and multi-stage transfer learning for the classification and localization of dysplasia in Barrett's esophagus. In: Suk, H.-I., Liu, M., Yan, P., Lian, C. (eds.) MLMI 2019. LNCS, vol. 11861, pp. 169–177. Springer, Cham (2019). https://doi.org/10.1007/978-3-030-32692-0_20
8. Rosenkrantz, A.B., Lim, R.P., Haghighi, M., Somberg, M.B., Babb, J.S., Taneja, S.S.: Comparison of interreader reproducibility of the prostate imaging reporting and data system and Likert scales for evaluation of multiparametric prostate MRI. Am. J. Roentgenol. **201**(4), W612–W618 (2013)
9. van der Sommen, F., et al.: Computer-aided detection of early neoplastic lesions in Barrett's esophagus. Endoscopy **48**(07), 617–624 (2016)
10. Watadani, T., et al.: Interobserver variability in the CT assessment of honeycombing in the lungs. Radiology **266**(3), 936–944 (2013)
11. Zhang, L., et al.: Disentangling human error from ground truth in segmentation of medical images. In: Larochelle, H., Ranzato, M., Hadsell, R., Balcan, M., Lin, H. (eds.) Advances in Neural Information Processing Systems. vol. 33, pp. 15750–15762. Curran Associates, Inc. (2020)

Improved Pancreatic Tumor Detection by Utilizing Clinically-Relevant Secondary Features

Christiaan G. A. Viviers[1]([✉]), Mark Ramaekers[2], Peter H. N. de With[1], Dimitrios Mavroeidis[3], Joost Nederend[2], Misha Luyer[2], and Fons van der Sommen[1]

[1] Eindhoven University of Technology, 5612 AZ Eindhoven, The Netherlands
`c.g.a.viviers@tue.nl`
[2] Catharina Ziekenhuis, 5623 EJ Eindhoven, The Netherlands
[3] Philips Research, 5656 AE Eindhoven, The Netherlands

Abstract. Pancreatic cancer is one of the global leading causes of cancer-related deaths. Despite the success of Deep Learning in computer-aided diagnosis and detection (CAD) methods, little attention has been paid to the detection of Pancreatic Cancer. We propose a method for detecting pancreatic tumor that utilizes clinically-relevant features in the surrounding anatomical structures, thereby better aiming to exploit the radiologist's knowledge compared to other, conventional deep learning approaches. To this end, we collect a new dataset consisting of 99 cases with pancreatic ductal adenocarcinoma (PDAC) and 97 control cases without any pancreatic tumor. Due to the growth pattern of pancreatic cancer, the tumor may not be always visible as a hypodense lesion, therefore experts refer to the visibility of secondary external features that may indicate the presence of the tumor. We propose a method based on a U-Net-like Deep CNN that exploits the following external secondary features: the pancreatic duct, common bile duct and the pancreas, along with a processed CT scan. Using these features, the model segments the pancreatic tumor if it is present. This segmentation for classification and localization approach achieves a performance of 99% sensitivity (one case missed) and 99% specificity, which realizes a 5% increase in sensitivity over the previous state-of-the-art method. The model additionally provides location information with reasonable accuracy and a shorter inference time compared to previous PDAC detection methods. These results offer a significant performance improvement and highlight the importance of incorporating the knowledge of the clinical expert when developing novel CAD methods.

Keywords: Pancreatic cancer · Tumor segmentation · CNN

1 Introduction

Pancreatic cancer is one of the leading causes of cancer-related deaths worldwide with a dismal prognosis and an overall 5-year survival rate of 9% [10]. Due to late

© The Author(s), under exclusive license to Springer Nature Switzerland AG 2022
S. Ali et al. (Eds.): CaPTion 2022, LNCS 13581, pp. 139–148, 2022.
https://doi.org/10.1007/978-3-031-17979-2_14

recognition, most patients advance to late stages of the disease or even metastases. Pancreatic tumor detection using CT imaging is considered to be the gold standard for the detection of pancreatic cancer [13]. The obtained accuracies of pancreatic ductal adenocarcinoma (PDAC) detection using CT imaging or other radiological imaging techniques largely depends on radiological expertise. Lack of such expertise may result in delayed recognition, which is problematic since only 20% of patients at the time of diagnosis are eligible for resection [4]. Therefore, early detection of pancreatic cancer holds significant promise by enabling surgical treatment and improving treatment outcomes.

Initial diagnosis of pancreatic tumors through CT imaging maintains acceptable sensitivity measures of around 90% for pancreatic cancer diagnosis [7]. In general, pancreatic tumors appear hypodense compared to normal pancreatic parenchyma. However, indeterminate CT findings such as small tumor size, growth pattern, iso-attenuating pancreatic cancer and the difficulty in differentiating from chronic pancreatitis, can make accurate delineation of viable tumor tissue a troublesome task [1]. In addition, pancreatic cancer often causes nonspecific symptoms prior to developing into an advanced stage. Therefore, it is important to identify secondary features which might indicate disease to improve early detection of PDAC. Computer-aided diagnosis and computer-aided detection (CAD) techniques hold great promise in enabling the early detection of PDAC. Such a tool allows for expert knowledge to be captured and shared, which can be used when the patient is first screened for the disease. Deep learning-based CAD methods have achieved impressive results in recent years. For these methods to be successfully adopted in the clinical environment, it is necessary to provide more than the standard "black-box" machine learning model [6,9]. For clinical acceptance of this technology, on top of high detection accuracies, it is essential to provide additional insights into the model's operation.

In this research, we propose a PDAC segmentation model that utilizes the same visual cues in the surrounding anatomy that experts use when looking for the presence of PDAC. This focus and way of working is to maximally leverage easily accessible external information and fully exploit clinical expertise, to ultimately optimize classification and localization performance. Since we start from the radiologists' reasoning, our method becomes clinically meaningful. For instance, a clinician pays close attention to pancreatic ductal size as a large (potentially dilated) duct could be indicative of tumor. Compared to normal pancreatic tissue in a CT scan, pancreatic cancer appears less visible as an ill-defined mass. It enhances poorly and is hypodense between 75% and 90% of arterial phase CT cases. For this reason, experts utilize secondary features which may be predictive of pancreatic cancer. These include, but are not limited to: ductal dilatation, hypo-attenuation, ductal interuption, distal pancreatic atrophy, pancreatic contour anomalies and common bile duct dilation. For a detailed description of these indicators, we refer to the work by Zhang et al. [15].

As these secondary features offer crucial information to experts during analysis, we hypothesize that a Deep Learning-based CAD method could also explicitly leverage this information. As such, we enrich the input of a 3D U-Net [3] segmentation model with an indication of the external secondary features and observe state-of-the-art results in PDAC detection.

2   Related Work on PDAC Detection

Deep learning has rapidly advanced the development of CAD methods across various domains. Invaluable research towards automated PDAC detection has also been conducted. Recently, both Liu et al. [8] and Si et al. [11] implemented a patch-based PDAC classification of CT volumes. These classification methods show high accuracy, but clinicians require more interpretable results, such as an indication to the tumor area. This will enable further insights about the tumor for potential follow-up tasks. To this end, a semantic segmentation approach is more suitable, since it additionally provides tumor localization. A Multi-Scale Coarse-to-Fine Segmentation method is proposed by Zhu et al. [16] that makes use of three U-Net-like segmentation models at different resolutions in a segmentation-for-classification approach. The output of the three networks are combined using a connected component graph between the adjacent tumor-positive voxels. Finally, false positive components from the graph are pruned and the tumor voxels are selected based on empirically selected thresholds. We refer to the work by Zhu et al. as previous state-of-the-art with a sensitivity of 94.1% and a specificity of 98.5%. Similarly, Alves et al. [2] proposed a similar segmentation-for-classification approach that makes use of four nnU-Net-based [5] models to ultimately detect the presence of the tumor. Although these methods achieve impressive results, it is engineering solutions that lacks transparency, sufficient motivation from a clinical perspective, in addition to suffering from long inference times. We propose a more intuitive, clinically-motivated method for PDAC detection. The proposed approach utilizes clinically-relevant cues to realize state-of-the-art detection scores while significantly simplify the network architecture, making it more suitable for deployment in medical centres.

3   Methods

3.1   Data Collection

In our retrospective single-center research study, we collected contrast-enhanced CT images of 97 control cases and 99 cases with pancreatic ductal adenocarcinoma (PDAC) located in the pancreatic head from the Catharina Hospital Eindhoven. Patients aged 18 years or above who underwent surgical treatment at the Catharina Hospital Eindhoven for pancreatic head cancer, were eligible when both a surgical report and a complete pathology report were available. All CT-scans were manually annotated in preparation for being used in this research. Relevant anatomical structures (tumor, pancreas, pancreatic duct and common bile duct) were annotated by a surgical resident and supervised by an expert abdominal radiologist using IntelliSpace Portal (Software package available from Philips Healthcare, The Netherlands). Patients in the control group

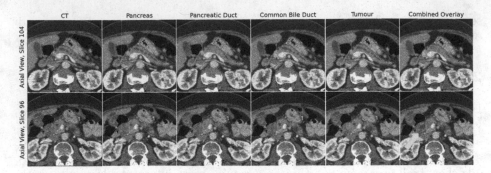

Fig. 1. Two slices from a case showcasing the involvement of the different structures and dilated ducts caused by the tumor blockage. These features are indicative of the tumor presence and its location. The bottom right image shows the involvement between the tumor and pancreas in dark red. (Color figure online)

were derived from a previous randomized control trial in which patients with esophageal cancer were included. These patients all had a CT scan as a preoperative work-up. The external secondary features play an important role in the expert radiologist's decision-making w.r.t. tumor presence, size and location. As such, much annotation effort was spent on not only the tumor, but also these indicative features. Two important factors that arose during this process were (1) how to annotate some of structures that belong to the same organ (pancreatic duct inside the pancreas) and (2) defining cases where there is a gradual transition from one structure to the other. The latter occurs when the common bile duct enters the pancreas, but importantly, also with the tumor itself. We decided that each structure should be annotated and stored separately to attain maximum information. However, this implied that CT voxels could potentially belong to multiple structures simultaneously. Figure 1 depicts an example case and corresponding ground-truth annotations. The last image in the bottom row shows the overlap between the pancreas and the tumor and the pancreatic duct in the pancreas.

As part of the Medical Decathlon [12] (MD), Task 07 involves the segmentation of the pancreas and pancreatic masses (intraductal papillary neoplasms, pancreatic neuroendocrine tumors, or pancreatic ductal adenocarcinoma). This dataset consists of patients with often well developed late-stage disease. As a result, there is a high proportion of large tumors and easily detectable cysts in this dataset. In addition, due to the extensive disease and associated symptoms, many cases contain metal stents, which could induce a bias in a learning algorithm. To the best of our knowledge, this is the only publicly available dataset that aims to detect pancreatic cancer, and although very valuable, it is still a step away from being an ideal dataset for training a deep learning-based CAD system for detection of PDAC. To provide some insight into how our approach competes against other methods on this public benchmark, we have supplemented 10% of

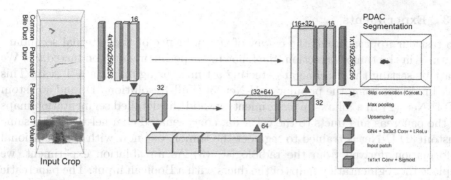

Fig. 2. Diagram of the 3D U-Net used for tumor segmentation in abdominal CT scans, provided with detailed external secondary feature segmentation maps.

this dataset's training set (28 cases) with suspected adenocarcinoma in the pancreatic head with separate annotations for the pancreatic duct, common bile duct, the full pancreas (unobstructed by the tumor) and the tumour. This subset will be used as an unseen test set in our experiments[1].

3.2 Segmentation Model for Classification and Localization

To provide clinicians with the necessary assistance in the early diagnosis of pancreatic cancer, an artificial intelligence system that is capable of accurately detecting PDAC in a clinically-interpretable way needs to be deployed. The system does a first screening, filtering out a large number of normal cases and preserving the cases with an indication of cancer. A CAD system that uses the same indicative features as the clinician could potentially be more explainable as it will provide additional insights as to how the result was derived. This could inspire more trust in the CAD system by the clinician, since it uses the same clinical way of working. As a result, the tumor is more likely to be detected earlier, leading to a better outcome. Bearing this in mind, we develop a segmentation for classification and localization method. We use a standard 3D U-Net (Fig. 2) that takes the segmentation maps capturing the external indicators as input and segments the tumor in the CT volume. These external secondary features are used by expert radiologists to identify and localize the tumor, but are also much more easily obtainable and can be identified in a non-expert setting with minimal effort. In practice, these can also be obtained by a prior segmentation model to streamline the process even further. Taking the difficulties related to accurate segmentation of pancreatic tumor into account (even for an expert radiologist), our objective is not to acquire a detailed segmentation map. Instead, we aim for an indication of where the tumor could be located. A detailed segmentation network or radiologist can then initiate follow-up work.

---

[1] Newly annotated data: https://github.com/cviviers/3D_UNetSecondaryFeatures.

3.3   Experiments

To test our approach and the extent of the influence of the external secondary features in the tumor detection, the following experiments are conducted. (1) We start by setting the baseline at detecting a tumor using only the CT scan. This baseline is set using the popular nnU-Net [5] (Full-Resolution 3D) and a custom 3D U-Net. (2) In a follow-up experiment, we add the detailed segmentation maps of the pancreas and ducts to the CT scan, concatenated channel-wise. The same custom 3D U-Net is trained to segment the tumor, but now with this additional information derived from the radiologist. (3) As an ablation experiment, we replace the segmentation maps of the ducts with a Boolean input. The pancreatic and common bile ductal 3D volumes are replaced with unity values if they are dilated or not. (4) Finally, we apply the models (using the CT scan and detailed segmentation maps), trained and validated on the three data folds of our dataset, to the Medical Decathlon Dataset as test set.

3.4   Data Preparation and Training Details

The radiologist starts the investigation for a tumor by localizing the pancreas in the CT scan. Once the pancreas have been located, the radiologist slides through scans looking for the various aforementioned indicative secondary features of the cancer. As such, we preprocess our data according to this expert's way of working. We derive a detailed segmentation map of the pancreas, pancreatic duct and common bile duct from the abdominal CT scan and use it as the secondary features. In practice, this is performed by a prior segmentation model, but since this is outside the scope of this paper, we use the ground-truth detailed segmentations provided by the expert radiologist. We crop the CT scan and corresponding labels, centered around the pancreas' center of mass. The crop is shaped within the dimensions [192, 256, 256] in the z, x, y-axes, respectively. Additional resampling and normalization is performed as described in the work by Isensee *et al.* [5] prior to cropping. We stack the CT scan, pancreas and two ducts channel-wise along a 4th dimension in preparation for training. Our final dataset is: $D = \{(\mathbf{X}_1, \mathbf{Y}_1), ...(\mathbf{X}_N, \mathbf{Y}_N)\}$, with N being the dataset size, $\mathbf{X}_n \in \mathbb{R}^{C \times Z \times W \times H}$ is the 4D volume of input data and $\mathbf{Y}_n \in \mathbb{R}^{Z \times W \times H}$ is the 3D tumor segmentation map.

In our implementation, we perform threefold cross-validation using a random 70/30% training and validation split and report results on the validation sets and the MD dataset as test set. The custom 3D U-Net is implemented in PyTorch and extends on the work by Wolny *et al.* [14]. During training we only employ a cross-entropy loss, a batch size of 2, an Adam optimizer with an initial learning rate of $1 \cdot 10^{-4}$ and a weight decay of $1 \cdot 10^{-5}$. We use extensive data augmentation, consisting of random flipping, random rotation, elastic deformation, contrast adjustment, and additive Gaussian and Poisson Noise. In all our experiments, the same crops, hyperparameters and augmentation techniques are used, with a hardware configuration based on a TITAN RTX GPU[2].

---
[2] Commercially available from Nvidia Corp., CA, USA.

4   Results and Discussion

The experimental results are listed in Table 1. In all cases, the model outputs are binarized (standard threshold setting of 0.5) and converted to segmentation maps. For the classification metric, if the resulting segmentation prediction overlaps with the ground-truth tumor label, even partially, we consider it a true positive prediction. If there is a yes (no) prediction and a tumor without overlap, it is a false positive (negative). In the case there is no prediction whatsoever and the tumor is absent, we consider it a true negative. The sensitivity, specificity and average Dice (across all the tumor-positive cases) on the validation sets are reported. We also show the results of the model using the full input (CT scan and detailed segmentation maps) and applied to the test MD dataset.

Results on Baseline: The nnU-Net and our 3D U-Net showcase similar performance when trained using only the CT scan as training data. In both models, the network eagerly tries to segment the tumor, even when the tumor is absent, resulting in a low specificity.

Adding Binary Ducts: The segmentation performance does not improve when the model is trained using the additional *binary* labels, indicating the presence of dilated ducts. Hence, duct dilation alone is not a tumor-deciding factor and has to be combined with indicative features from the hypodense tumor region.

Detailed Segmentation Maps: The model is provided with the detailed segmentation maps of the ducts and pancreas, along with the CT scan. We observe that the model can learn the connection between these indicative features and the presence of a tumor. The model correctly predicts tumor with an overlap of the ground-truth segmentation in the majority of cases. In a single case, tumor is predicted to be at a different location than the label (False Positive). We observe a lower Dice score compared to the models with only CT scans as input. These baseline models maximally predict tumors in most cases. This results in a higher Dice score when there is a tumor factually present, at the expense of a large number of false positive predictions. When increasing sensitivity, the model logically locates more tumors, albeit some with low Dice scores.

Table 1. Results obtained with the nnU-Net and 3D U-Net with different input channel information. Given the limited amount of data, numbers are constrained to two decimals.

Data input	Model	Sensitivity	Specificity	Dice
CT only	nnU-Net	0.92 ± 0.02	0.27 ± 0.16	0.42 ± 0.04
CT only	3D U-Net	0.98 ± 0.03	0.11 ± 0.10	0.40 ± 0.07
Binary ducts	3D U-Net	0.83 ± 0.24	0.19 ± 0.06	0.16 ± 0.04
Full	3D U-Net	1.00 ± 0.00	0.99 ± 0.02	0.31 ± 0.07
Test MD - Full	3D U-Net	0.99 ± 0.02	N/A	0.31 ± 0.05

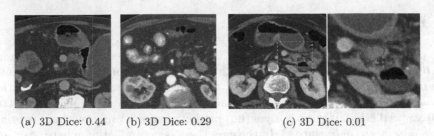

(a) 3D Dice: 0.44   (b) 3D Dice: 0.29   (c) 3D Dice: 0.01

Fig. 3. Segmentation performance from three different cases. An example of a low-performing segmentation is visualized (3c). The figure is best viewed in color.

Test Set (MD) Experiments: We observe very similar impressive performance when the model is applied to the MD dataset as test set. In two of the three folds, the models showcase 100% sensitivity, with no tumor missed. The model from the third fold missed the tumor in one of the cases and made no prediction whatsoever (False Negative). The same tumor was predicted with a relatively high Dice score of 0.40 and 0.26 in the other two models. Averaging the sensitivity across the three models, 100%, 100% and 96.43%, explains the $99 \pm 2\%$ sensitivity at the bottom of the table (check remark on accuracy in table caption). Example predictions on the test set can be seen in Fig. 3. Note that the aim of this study is not to achieve maximum segmentation accuracy, but rather develop a more effective, clinically-relevant and efficient method for tumor detection. Inference time using this method is 0.33 s on a RTX 2080 Ti.

Limitations: The secondary features used in this work and provided as external input are acquired from the same CT scan. One would expect that a CNN would be able to extract these embedded spatial features and discover the causality between these features and the presence of the tumor. Unfortunately, this expectation is not valid. Future work should investigate these underlying causal factors and how to enable a CNN to exploit this available information.

5   Conclusion

Despite the eminent success of deep learning networks, even for detection of PDAC, the method presented in this work demonstrates that external tumor-indicative features can significantly boost CAD performance. We optimize a segmentation for classification and localization approach, by adding the easily obtainable and clinically valuable external secondary features used by the radiologist, to considerably improve segmentation performance. The proposed approach consists of a 3D U-Net that takes the CT scan, along with a segmentation map of the pancreas, pancreatic duct and common bile duct as input, in order to finally segment the pancreatic tumor. By integrating these indicative secondary features into the detection process, the proposed method achieves a sensitivity of $99 \pm 2\%$ (one cased missed), yielding 5% gain over the previous state-of-the-art method. The proposed method also achieves a specificity of 99% and ultimately requires no sacrifice of specificity in favor of sensitivity. In addition, the method provides further insights into the tumor location and obtain similar segmentation scores on prospectively collected and the Medical Decathlon data. Generally, this research reveals the important value of explicitly including clinical knowledge into the detection model. We suggest that future CAD methods integrate higher orders of feature information, particularly valuable clinical features, into their domain-specific problem to improve performance when such information can be identified. This method paves the way for equipping clinicians with the necessary tools to enable early PDAC detection, with the aim to ultimately improve patient care.

References

1. Ahn, S.S., et al.: Indicative findings of pancreatic cancer in prediagnostic CT. Eur. Radiol. **19**(10), 2448–2455 (2009)
2. Alves, N., Schuurmans, M., Litjens, G., Bosma, J.S., Hermans, J., Huisman, H.: Fully automatic deep learning framework for pancreatic ductal adenocarcinoma detection on computed tomography. Cancers **14**(2), 376 (2022)
3. Çiçek, Ö., Abdulkadir, A., Lienkamp, S.S., Brox, T., Ronneberger, O.: 3D U-Net: learning dense volumetric segmentation from sparse annotation. In: Ourselin, S., Joskowicz, L., Sabuncu, M.R., Unal, G., Wells, W. (eds.) MICCAI 2016. LNCS, vol. 9901, pp. 424–432. Springer, Cham (2016). https://doi.org/10.1007/978-3-319-46723-8_49
4. Hidalgo, M.: Pancreatic cancer. N. Engl. J. Med. **362**(17), 1605–1617 (2010)
5. Isensee, F., Jaeger, P.F., Kohl, S.A.A., Petersen, J., Maier-Hein, K.H.: NNU-Net: a self-configuring method for deep learning-based biomedical image segmentation. Nat. Methods **18**(2), 203–211 (2021)
6. Kriegsmann, M., et al.: Deep learning in pancreatic tissue: identification of anatomical structures, pancreatic intraepithelial neoplasia, and ductal adenocarcinoma. Int. J. Mol. Sci. **22**(10), 5385 (2021)
7. Lee, E.S., Lee, J.M.: Imaging diagnosis of pancreatic cancer: a state-of-the-art review. World J. Gastroenterol. **20**(24), 7864–7877 (2014)
8. Liu, K.L., et al.: Deep learning to distinguish pancreatic cancer tissue from noncancerous pancreatic tissue: a retrospective study with cross-racial external validation. Lancet Digital Health **2**(6), e303–e313 (2020)
9. Petch, J., Di, S., Nelson, W.: Opening the black box: the promise and limitations of explainable machine learning in cardiology. Can. J. Cardiol. **38**(2), 204–213 (2022)
10. Rahib, L., et al.: Projecting cancer incidence and deaths to 2030: the unexpected burden of thyroid, liver, and pancreas cancers in the united states. Cancer Res. **74**(11), 2913–2921 (2014)
11. Si, K., et al.: Fully end-to-end deep-learning-based diagnosis of pancreatic tumors. Theranostics **11**(4), 1982–1990 (2021)
12. Simpson, A.L., et al.: A large annotated medical image dataset for the development and evaluation of segmentation algorithms (2019)
13. Treadwell, J.R., et al.: Imaging tests for the diagnosis and staging of pancreatic adenocarcinoma: a meta-analysis. Pancreas **45**(6), 789–795 (2016)
14. Wolny, A., et al.: Accurate and versatile 3d segmentation of plant tissues at cellular resolution. eLife. **9**, e57613 (2020)

15. Zhang, L., Sanagapalli, S., Stoita, A.: Challenges in diagnosis of pancreatic cancer. World J. Gastroenterol. **24**(19), 2047–2060 (2018)
16. Zhu, Z., Xia, Y., Xie, L., Fishman, E.K., Yuille, A.L.: Multi-scale coarse-to-fine segmentation for screening pancreatic ductal adenocarcinoma. In: Shen, D., et al. (eds.) MICCAI 2019. LNCS, vol. 11769, pp. 3–12. Springer, Cham (2019). https://doi.org/10.1007/978-3-030-32226-7_1

Strategising Template-Guided Needle Placement for MR-targeted Prostate Biopsy

Iani JMB Gayo[1,2,3](✉), Shaheer U. Saeed[1,2,3], Dean C. Barratt[1,2,3], Matthew J. Clarkson[1,2,3], and Yipeng Hu[1,2,3]

[1] Department of Medical Physics and Biomedical Engineering, University College London, London, UK
iani.gayo.20@ucl.ac.uk
[2] Wellcome/EPSRC Centre for Interventional and Surgical Sciences, University College London, London, UK
[3] Centre for Medical Image Computing, University College London, London, UK

Abstract. Clinically significant prostate cancer has a better chance to be sampled during ultrasound-guided biopsy procedures, if suspected lesions found in pre-operative magnetic resonance (MR) images are used as targets. However, the diagnostic accuracy of the biopsy procedure is limited by the operator-dependent skills and experience in sampling the targets, a sequential decision making process that involves navigating an ultrasound probe and placing a series of sampling needles for potentially multiple targets. This work aims to learn a reinforcement learning (RL) policy that optimises the actions of continuous positioning of 2D ultrasound views and biopsy needles with respect to a guiding template, such that the MR targets can be sampled efficiently and sufficiently. We first formulate the task as a Markov decision process (MDP) and construct an environment that allows the targeting actions to be performed virtually for individual patients, based on their anatomy and lesions derived from MR images. A patient-specific policy can thus be optimised, before each biopsy procedure, by rewarding positive sampling in the MDP environment. Experiment results from fifty four prostate cancer patients show that the proposed RL-learned policies obtained a mean hit rate (HR) of 93% and an average cancer core length (CCL) of 11 mm, which compared favourably to two alternative baseline strategies designed by humans, without hand-engineered rewards that directly maximise these clinically relevant metrics. Perhaps more interestingly, it is found that the RL agents learned strategies that were adaptive to the lesion size, where spread of the needles was prioritised for smaller lesions. Such a strategy has not been previously reported or commonly adopted in clinical practice, but led to an overall superior targeting performance, achieving higher HR (93% vs 76%) and measured CCL (11.0 mm vs 9.8 mm) when compared with intuitively designed strategies.

Keywords: Reinforcement learning · Prostate cancer · Targeted biopsy · Planning

© The Author(s), under exclusive license to Springer Nature Switzerland AG 2022
S. Ali et al. (Eds.): CaPTion 2022, LNCS 13581, pp. 149–158, 2022.
https://doi.org/10.1007/978-3-031-17979-2_15

1   Introduction

Recent development in multiparametric MR imaging (mpMRI) techniques provides a means of noninasive localisation of suspected prostate cancer [1], which enables clinicians to target these lesions during the follow-up ultrasound-guided biopsy for further histopathology confirmation. This MR-targeted approach has been shown to reduce both the false positive and false negative detection, compared to previously adopted random biopsy [1,2], and subsequently motivated research in developing multimodal MR-to-ultrasound image registration [3].

Needle sampling of the MR-identified targets, with or without registration errors, can still be a challenging and arguably overlooked task. Operator expertise was found to be an important predictor in detecting clinically significant prostate cancer [4]. Planning strategies is important for navigating the ultrasound probe, to better observe the targets with respect to imaging, and for manual needle positioning. In transperineal biopsy, the introduction of brachytherapy templates (See Fig. 1 for an example) helps the needle deployment - a procedure that is of interest in this study, but choice between 13×13 grid positions remains a subjective decision. For example, a common clinical practice aims at the target centre, but it has been shown to yield an insufficient sampling of the heterogeneous cancer [5] and possibly an inferior diagnostic accuracy in terms of disease-representative grading [6], compared with more spread needle placement. The design of an optimum strategy is further complicated by the need of multiple needles for individual targets, for maximising the hit rate, and the multifoci nature of prostate cancer, which requires repeated sampling of one or more targets.

To the best of our knowledge, there has not been any computer-assisted sampling strategy that takes into account the previous needle deployment(s) or quantitatively optimises patient-and-target-specific needle distribution. In summary, improving the targeting strategy may help reduce the significant false negative rate found in MR-targeted biopsy (reported being as high as 13% [7]), and hence improves the chance of early cancer detection for patients.

Reinforcement learning (RL) has been proposed for medical image analysis tasks [8] such as plane finding [9], and for surgical planning in hysterectomy [10] and orthopaedic operations [11]. It has also been used for needle path planning in minimally-invasive robotic surgery [12]. It is its ability to learn intelligent policies for sequential decision making that provides a potential solution to problems without requiring direct supervision for each action, a common constraint in developing machine learning methods for skill-demanding surgical and intervention applications. This makes RL suitable for finding an optimal targeting strategy, which requires complex decisions for which there is no established best method.

In this study, we investigate the feasibility of using RL to plan patient-specific needle sampling strategies, optimised in pre-operative MR-derived RL environments. We present experimental results based on clinical data from prostate cancer patients and compare the proposed method, using a set of clinically important metrics, to baseline strategies that are designed by human intuition and an interactive targeting experiment performed by two observers. We conclude by reporting a set of interesting observations that demonstrate the benefit of using the proposed RL-learned patient-specific strategies. These indeed adapted effectively to individual procedures and varying targets, achieving consistent hit rates with less variance for both smaller and larger lesions.

2 Method

The agent-environment interactions are modelled as a Markov decision process (MDP), and summarised as a 4-tuple $\langle \mathcal{S}, \mathcal{A}, r, p \rangle$, where $\mathcal{S}$ and $\mathcal{A}$ are the state and action spaces consisting of all possible observed states as input and actions as output for the agent, respectively. $r : \mathcal{S} \times \mathcal{A} \rightarrow \mathbb{R}$ is the reward function which maps state-action pairs to a real value. The state transition distribution is defined by $p : \mathcal{S} \times \mathcal{S} \times \mathcal{A} \rightarrow [0, 1]$ which denotes probability of transitioning to the next state, given the current state-action pair. In this section, we develop an environment for template-guided biopsy sampling of the cancer targets, the MDP components and the policy learning strategy.

2.1 Patient-specific Prostate MR-derived Biopsy Environment

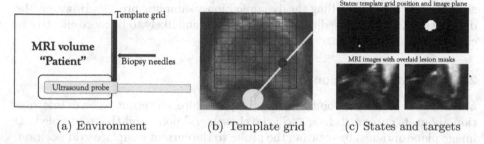

(a) Environment   (b) Template grid   (c) States and targets

Fig. 1. Simulated biopsy procedure environment. (a) Placement of ultrasound probe and template grid within the MRI volume. (b) Visualisation of ultrasound probe rotation which is always aligned with the chosen template grid position. (c) Examples of (top) states and (bottom) overlaid MR-identified targets

The environment is illustrated in Fig. 1a for the targeted biopsy procedures, where virtual biopsy needles are inserted through the perineum via a brachytherapy template grid consisting of 13×13 holes that are 5 mm apart. Other needle-based treatments such as cryotherapy, brachytherapy and radiofrequency ablation may also be applicable but are not discussed further in this paper. The position of the transrectal ultrasound probe is approximated within the rectum directly underneath the prostate gland, with a fixed distance to the template grid such that the top of the probe is aligned with the lower side of the template, as shown in the Fig. 1a. Both anatomical and pathological structures can be sampled, at any sagittal ultrasound imaging plane given an arbitrary angle, as illustrated in Fig. 1b.

We summarised a number of considerations in designing and constructing the adopted biopsy environment as follows. 1) The prostate gland from each MR volume, the MR-identified targets (as in Fig. 1c), and key landmarks such as position of the rectum are all manually segmented from individual patients to construct the biopsy environment. Automated methods for segmenting these regions of interest have been proposed, e.g. [13,14]. 2) Binary segmentation are provided as observations for the RL agents, as opposed to ultrasound image intensities, which are neither available during planning nor straightforward to synthesize from MR images. We argue that the use of binary representation would be more robust to train the RL agents and the resulting methods are more likely to generalise to different procedures and planning MR images, especially given the existing MR and ultrasound segmentation and registration algorithms described above. 3) Uncertainty in MR-to-ultrasound registration should be added to the segmented regions, together with other potential errors in localising these regions such as observer variability in manual segmentation used in this work. We would like to point out that, however, these localisation errors are unlikely to be independent to each other and the dependency of RL model generalisability on how precisely these need to be modelled remain open research questions. Section 3 discusses further details adopted in our experiments. 4) In the presented experiments, we focus on targeting the index lesions, those are of largest volumes in each case, to provide first results that show the efficacy of modelling the dynamic biopsy sampling process. However, the described MDP should be directly applicable for and likely to be more effective in cases with multiple lesions.

2.2   The MDP Components

State - At a given time point t during the procedure, the agent receives information about its current state $s_t \in \mathcal{S}$: the chosen grid point and the re-sampled 2D image plane obtained by rotating the probe to the current template grid position, as in Fig. 1c. This current position is determined by the previous action. This is to test the scenario with least assumptions, where the overall 3D anatomical and pathological information may be corrupted or unreliable due to intra-procedural uncertainties from patient movement and outdated registration.

Actions - The agent takes actions $a_t \in \mathcal{A}$ which modify its position on the template grid. These actions are relative to the current position of the agent (i, j) and are defined as $a_t = (\delta_i, \delta_j)$ such that the new position is given by $(i + \delta_i, j + \delta_j)$, where $\delta_i, \delta_j \in [-15, 15]$. By formulating this relative grid-moving action, we consider the biopsy needle is positioned on the image plane, with an insertion depth that overlaps the needle centre and the centre of the observed 2D target, subject to small predefined positioning errors in each direction. These are commonly adopted practice though not strictly enforced, and we found that increasing the flexibility by independently positioning the ultrasound probe and needle may unnecessarily make the training difficult to converge.

Reward - The reward at the time t is computed based on the reward function $R_t = r(s_t, a_t)$, during training. The agent is rewarded positively if the fired needle obtains samples of the lesion. A penalty is given when chosen grid positions are outside of the prostate, to avoid hitting surrounding critical structures. From initial experiments, it was found that a greater reward of +5 lead to a faster convergence during training, encouraging the agent to hit the lesions, whilst a penalty of −1 was enough to deter the agent from firing outside the prostate gland. Reward shaping is also introduced to guide the agent towards the lesion, thereby speeding up the learning process. Similar to [9], a sign function Sgn of the difference between $dist_{t-1}$ and $dist_t$ is computed, where $dist_t$ represents Euclidean distance between target centre and needle trajectory at time t.

$$r = \begin{cases} +5 & \text{if biopsy needle intersected with target} \\ -1 & \text{if biopsy needle placed outside prostate} \\ \text{Sgn}(dist_{t-1} - dist_t) & \text{otherwise} \end{cases} \quad (1)$$

2.3 Policy Learning

The navigation and sampling strategy is parameterised by a policy neural network π_θ, with parameters θ, quantifying the probability of performing action a_t given state s_t. Agent's actions then can be sampled from the policy, $a_t \sim \pi_\theta(\cdot|s_t)$. During the policy training, the accumulated reward $Q^{\pi_\theta}(s_t, a_t) = \sum_{k=0}^{T} \gamma^k R_{t+k}$ is maximised, where γ is a discount factor set to 0.9, to obtain the optimal policy π_{θ^*}, $\theta^* = \arg\max_\theta \mathbb{E}_{\pi_\theta}[Q^{\pi_\theta}(s_t, a_t)]$. With continuous actions, policy gradient (PG) and actor-critic (AC) algorithms can thus be adopted for the optimisation.

3 Experiments

Data Set - The T2-weighted MR images and their segmentation were acquired from 54 prostate cancer patients. These were obtained as part of clinical trials, PROMIS [1] and SmartTarget [7], where patients underwent ultrasound-guided minimally invasive needle biopsies and focal therapy procedures.

RL Algorithm Implementation - An agent was trained for each patient individually using the Stable Baselines implementation of PPO [15]. Each agent was trained for 120,000 episodes and a model was selected with the highest average reward after 10 episodes. Each episode was limited to a maximum of 15 time steps, but can terminate early if any 5 needles hit the lesion. At each episode the agent is initialised at random starting positions on the template grid. Based on estimated registration error reported previously [3], random localisation error was added in the observed states, equivalent to a Gaussian noise with a standard deviation of 1.73 mm in each of the x, y and z coordinates, or a mean distance error of 3 mm. The PPO algorithm [16] was used in reported results, as it guarantees a monotonic reward improvement and stability of training. However, we also report a lack of substantial difference in performance to other tested algorithms, including DDPG [17] and SAC [18]. The policy network was based on ResNet18 [19] architecture, with an additional fully-connected layer for a linear output.

154   I. J. Gayo et al.

An Adam optimiser was used, with a learning rate of 0.0001. It could be of future interest to compare further network architectures and PG/AC algorithms on the proposed RL problem, but is considered beyond the scope of current work.

Biopsy Performance Metrics - To quantitatively assess target sampling performance, three biopsy-specific metrics are used in this study: hit rate (HR), cancer core length (CCL) and needle area (NA). The HR is the number of needle samples that contains the target, i.e. positive samples, divided by the number of needles fired. Five needles are chosen to represent the maximum number typically used in targeted biopsy [20]. CCL is the total length (in mm) overlapping the target, i.e. the sampled target tissue, in individual needles. CCL $>=$ 6 mm often indicates clinical significance [21]. NA estimates the coverage of all fired needles in each episode, defined as the area of an approximating ellipse, $NA = \pi * std_x \times std_y$, where std_x and std_y are standard deviations in the needle navigating x-y plane, defined by the template grid position.

Baseline Strategies - Two strategies were compared with the proposed agent, designed to provide an estimate of what clinicians are likely to achieve in practice. For a fair comparison, the same observed targets, the states, and starting positions were used. Student's t-tests were used when comparison is made at a significance level of $\alpha = 0.05$, unless otherwise indicated. The first strategy *(Sweeping strategy)* adopted a simple sweeping of the biopsy needle together with the ultrasound probe, from left to right in a 5 mm interval. The target was sampled at the centre of the observed target, i.e. fired, once an image plane is encountered with a lesion. The second strategy *(Scouting strategy)* moves the virtual probe to scout all candidate positions that samples the target, before 5 random ones were selected as fired positions among these candidates. Inter- and intra-operator variance is one of the most important factors that impact the performance of a sampling strategy. For the baseline strategies, additional Gaussian noise was added to the chosen needle/probe positioning with a varying bias and a varying standard deviation (SD). Increasing the SD leads to higher uncertainty in placing the needles, while the bias indicates a targeting strategy that does not aim for the target centre, e.g. for avoiding empty cores or urethra. The experiments were repeated using different combinations of the two variables (each ranging between 0 to 10 mm), to test different strategies.

Interactive Experiments by Two Observers - Two human operators, one computer scientist and one biomedical imaging researcher, interacted with a custom-made interface that displays the current template grid position and image plane observed. They were asked to choose where to sample and how far to spread the needles. This is a simple interactive experiment to provide preliminary results that can be compared with the other described strategies.

4   Results

Learned Strategy Performance - From Table 1, the agent outperforms both baseline strategies in HR and CCL (both p-values < 0.005), but not in NA, with noticeably smaller variability in HR. Different levels of bias did not lead to significantly different targeting results, while higher SD increased the spread of needles, but reduced CCL and HR. In general, the results between the sweeping and scouting strategies were not found statistically different in CCL and NA. The scouting strategy resulted in an increased HR (p-value < 0.05) compared to sweeping, but does not outperform the agent which still obtains the highest HR (Table 1).

Table 1. Summary of biopsy performance from the RL agent (top row) and the sweeping and scouting strategies for different bias and SD combinations

Bias	SD	Baseline 1 (Sweeping)			Baseline 2 (Scouting)		
		CCL(mm)	HR(%)	NA(mm^2)	CCL(mm)	HR(%)	NA(mm^2)
Agent		11.13 ± 3.43	93.40 ± 11.44	22.14 ± 18.18	11.13 ± 3.43	93.40 ± 11.44	22.14 ± 18.18
0	0	7.95 ± 3.00	53.36 ± 35.59	23.67 ± 16.57	8.40 ± 2.65	61.10 ± 30.01	31.31 ± 28.53
0	5	7.45 ± 3.19	49.43 ± 34.10	56.81 ± 50.25	8.32 ± 3.04	55.19 ± 30.46	39.84 ± 20.48
0	10	4.89 ± 3.29	30.94 ± 28.71	108.00 ± 86.24	5.67 ± 4.32	43.70 ± 31.99	114.35 ± 71.20
5	0	8.90 ± 4.00	53.69 ± 28.44	36.53 ± 28.91	8.32 ± 3.00	65.19 ± 30.04	36.53 ± 28.91
5	5	7.88 ± 3.52	51.32 ± 32.94	60.44 ± 47.62	7.28 ± 3.17	54.48 ± 31.04	72.08 ± 40.89
5	10	6.06 ± 3.95	41.51 ± 31.58	86.15 ± 63.15	5.74 ± 4.41	44.82 ± 31.02	113.88 ± 79.60
10	0	7.29 ± 3.08	47.80 ± 34.16	27.47 ± 18.38	7.65 ± 3.05	54.81 ± 31.13	27.59 ± 22.95
10	5	7.14 ± 3.38	42.44 ± 28.00	52.78 ± 48.02	6.40 ± 3.64	50.74 ± 29.00	55.45 ± 42.36
10	10	4.74 ± 3.46	28.70 ± 27.22	106.17 ± 115.95	5.98 ± 3.42	47.04 ± 31.22	112.68 ± 95.85

Table 2. Obtained biopsy strategy metrics for the agent and two human observers

Observers	CCL (mm)	HR(%)	NA (mm^2)
Agent	11.13 ± 3.43	93.40 ± 11.44	22.14 ± 18.18
Observer 1	9.71 ± 3.78	66.30 ± 20.55	42.85 ± 23.36
Observer 2	9.83 ± 3.89	76.30 ± 19.50	64.90 ± 38.84

Interactive Experiments by Two Observers - The agent outperforms both observers in CCL (p-values = 0.020 & 0.040), but its NA values are more than double that of the agents, suggesting a potential trade-off between sampling coverage and precision. For HR, the agent outperforms both observers (p-value < 0.001), which demonstrates that the agent could achieve an overall comparable performance as human observers, with a significantly higher CCL.

Learned Strategy for Varying Target Sizes - From Fig. 2a and Table 3, we observe an interesting behaviour from the learned agent: the smaller the lesions, the larger the spread of the needles. At a volume threshold of 0.4 cc, the mean CCL and NA are statistically different for smaller and larger lesions (p-values = 0.002 & 0.040), whilst the difference in HR is not (p-value = 0.166). This result may seem counter-intuitive, as one would be cautious in spreading needles for a small target. However, the agent learned to distribute needles more widely for smaller lesions, attempting to maintain the hit rate, given the inevitable presence of target localisation uncertainty described in Sect. 2.1. Visual examples of the learned strategies are shown in Figs. 2b and 2c. This learned behaviour is interesting because a) it has not been observed previously, either in literature or in clinical practice. b) it improved the overall targeting performance compared to the target-size-agnostic baseline strategies and c) this could be suggested to urologists and interventional radiologists with or without the proposed RL assistance.

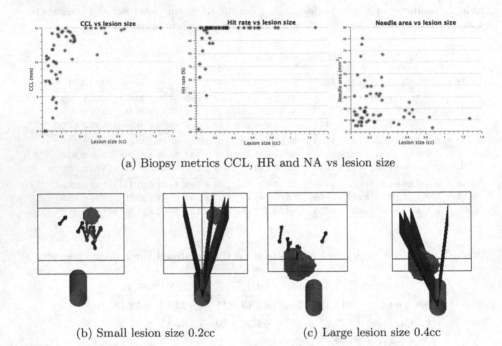

(a) Biopsy metrics CCL, HR and NA vs lesion size

(b) Small lesion size 0.2cc   (c) Large lesion size 0.4cc

Fig. 2. (a): Biopsy metrics vs lesion size. (b) and (c): Examples of learned strategies for different sized targets (red), represented by the needle positions (red sticks, brighter indicates later time steps) and observed ultrasound images in green. The bounding cube and cylinder represent the prostate volume and probe. (Color figure online)

Table 3. CCL, HR and NA for different lesion sizes using threshold size < 0.4cc

Lesion size	CCL (mm)	HR (%)	NA (mm^2)
Small lesions	10.26 ± 4.19	93.16 ± 16.03	25.26 ± 19.45
Large lesions	14.52 ± 1.18	100.00 ± 0.00	13.02 ± 6.25

5 Discussion and Conclusion

The results demonstrate that the RL agents are competitive in sampling MR-derived targets, achieving higher HR and CCL compared with intuitively devised strategies. Furthermore, the learned strategies adapted to patient-specific procedures and varying pathology. Similar HR was achieved for different sized lesions, by spreading the fired needles more for smaller lesions. Such behaviour has not been observed before, and could be suggested to clinicians for improved targeting performance. Assumptions, such as number of allowed needles and uncertainties in localisation/placement, have been made to facilitate the proposed pre-procedural planning. Overall, the improved targeting performance provides means in mitigating the cancer under-sampling and help timely diagnosis of a significant number of prostate cancer patients with current MR-targeted biopsy.

Acknowledgement. This work is supported by the EPSRC-funded UCL Centre for Doctoral Training in Intelligent, Integrated Imaging in Healthcare (i4health) [EP/S021930/1], EPSRC [EP/T029404/1], and NIHR funded Biomedical Research Centre at University College Hospital. This work is also supported by the International Alliance for Cancer Early Detection, an alliance between Cancer Research UK [C28070/A30912; C73666/A31378], Canary Center at Stanford University, the University of Cambridge, OHSU Knight Cancer Institute, University College London and the University of Manchester. This work is supported by the Wellcome/EPSRC Centre for Interventional and Surgical Sciences [203145Z/16/Z].

References

1. Ahmed, H.U., et al.: Diagnostic accuracy of multi-parametric MRI and TRultrasound biopsy in prostate cancer (PROMIS): a paired validating confirmatory study. The Lancet. **389**, 815–822 (2017). https://doi.org/10.1016/s0140-6736(16)32401-1
2. Simmons, L.A.M., et al.: The PICTURE study - Prostate Imaging (multiparametric MRI and Prostate HistoScanningTM) Compared to Transperineal Ultrasound guided biopsy for significant prostate cancer Risk Evaluation. Contemp. Clin. Trials **37**, 69–83 (2014). https://doi.org/10.1016/j.cct.2013.11.009
3. Hu, Y., et al.: Weakly-supervised convolutional neural networks for multimodal image registration. Med. Image Anal. **49**, 1–13 (2018). https://doi.org/10.1016/j.media.2018.07.002
4. Stabile, A.,et al.: Not all multiparametric magnetic resonance imaging-targeted biopsies are equal: the impact of the type of approach and operator expertise on the detection of clinically significant prostate cancer. Eur. Urol. Oncol. 1, 120–128 (2018). https://doi.org/10.1016/j.euo.2018.02.002
5. Calio, B.P., et al.: Spatial distribution of biopsy cores and the detection of intralesion pathologic heterogeneity. Therap. Adv. Urol. **11**, 1756287219842485 (2019). https://doi.org/10.1177/1756287219842485
6. Orczyk, C., et al.: MP38-07 Should we aim for the centre of an MRI prostate lesion? Correlation between mpMRI and 3-Dimensional 5mm transperineal prostate mapping biopsies from the PROMIS trial. J. Urol. **197**(4S), e486–e486 (2017)

7. Hamid, S., et al.: The smarttarget biopsy trial: a prospective, within-person randomised, blinded trial comparing the accuracy of visual-registration and magnetic resonance imaging/ultrasound image-fusion targeted biopsies for prostate cancer risk stratification. Eur. Urol. **75**, 733–740 (2019). https://doi.org/10.1016/j.eururo.2018.08.007

8. Zhou, S.K., Le, H.N., Luu, K., Nguyen, H.V., Ayache, N.: Deep reinforcement learning in medical imaging: a literature review. Med. Image Anal. **72**, 102193 (2021)

9. Alansary, A., et al.: Automatic View Planning with Multi-scale Deep Reinforcement Learning Agents. arXiv:1806.03228 [cs] (2018)

10. Sato, M., Koga, K., Fujii, T., Osuga, Y.: Can Reinforcement Learning Be Applied to Surgery? IntechOpen (2018)

11. Ackermann, J., et al.: A new approach to orthopedic surgery planning using deep reinforcement learning and simulation. In: de Bruijne, M., et al. (eds.) MICCAI 2021. LNCS, vol. 12904, pp. 540–549. Springer, Cham (2021). https://doi.org/10.1007/978-3-030-87202-1_52

12. Lee, Y., Tan, X., Chng, C.B., Chui, C.K: Simulation of Robot-Assisted Flexible Needle Insertion using Deep Q-Network (2019)

13. Aldoj, N., Biavati, F., Michallek, F., Stober, S., Dewey, M.: Automatic prostate and prostate zones segmentation of magnetic resonance images using DenseNet-like U-net. Sci. Rep. **10**, 14315 (2020). https://doi.org/10.1038/s41598-020-71080-0

14. Dai, Z., et al.: Segmentation of the prostatic gland and the intraprostatic lesions on multiparametic magnetic resonance imaging using mask region-based convolutional neural networks. Adv. Radiat. Oncol. **5**, 473–481 (2020). https://doi.org/10.1016/j.adro.2020.01.005

15. StableBaselines3. https://stable-baselines3.readthedocs.io/en/master/modules/ppo.html

16. Schulman, J., Wolski, F., Dhariwal, P., Radford, A., Klimov, O.: Proximal Policy Optimization Algorithms. https://arxiv.org/abs/1707.06347

17. Lillicrap, T., et al.: Continuous control with deep reinforcement learning (2019)

18. Haarnoja, T., Zhou, A., Abbeel, P., Levine, S.: Soft Actor-Critic: Off-Policy Maximum Entropy Deep Reinforcement Learning with a Stochastic Actor (2018)

19. He, K., Zhang, X., Ren, S., Sun, J.: Deep Residual Learning for Image Recognition. https://arxiv.org/abs/1512.03385

20. Song, G., et al.: How many targeted biopsy cores are needed for clinically significant prostate cancer detection during transperineal magnetic resonance imaging ultrasound fusion biopsy? J. Urol. **204**, 1202–1208 (2020). https://doi.org/10.1097/JU.0000000000001302

21. Ahmed, H.U., et al.: Characterizing clinically significant prostate cancer using template prostate mapping biopsy. J. Urol. **186**, 458–464 (2011). https://doi.org/10.1016/j.juro.2011.03.147

Semantic-Aware Registration with Weakly-Supervised Learning

Zhan Jin[1,2,3], Peng Xue[3], Yuyao Zhang[1], Xiaohuan Cao[2(✉)], and Dinggang Shen[1,3,4(✉)]

[1] School of Information Science and Technology, ShanghaiTech University, Shanghai, China
{jinzhan,xuepeng,zhangyy8,dgshen}@shanghaitech.edu.cn
[2] Shanghai United Imaging Intelligence Co. Ltd., Shanghai, China
xiaohuan.cao@uii-ai.com
[3] School of Biomedical Engineering, ShanghaiTech University, Shanghai 201210, China
[4] Shanghai Clinical Research and Trial Center, Shanghai 201210, China

Abstract. Medical image registration is a fundamental task for many clinical applications. Most deep learning-based image registrations methods have achieved brilliant performance owing to the incorporation of mask information. However, existing mask-guided registration methods only focus on volumetric registration inside the paired masks, ignoring the inherent attributes of the anatomical structure in other dimensions (e.g., smoothness of organ surface, connectivity of tubular structures, etc.). To address this problem, we proposed a novel semantic-aware registration network suitable for multi-organs registration with different anatomical structures. By analyzing the structural characteristics of various organs or tissues, semantic constraints are directly imposed on the deformation field from two geometrically meaningful dimensions (surface, contour) to preserve the topology of organs or tissues. To ensure the versatility of our proposed network, only a randomly selected segmentation mask is used for supervision at each iteration during training. Such a training strategy can satisfy the accurate registration of an individual organ at the inference stage. Experiments on a pelvic CT dataset with 112 subjects show that our method can achieve higher registration accuracy and preserve the anatomical structure more effectively than state-of-the-art methods.

1 Introduction

Deformable registration is crucial for medical image analysis tasks, such as image fusion, image-guided intervention, and radiation therapy. Over the past few years, deep learning has drawn much attention to solving image registration problems. Many deep learning-based registration methods have been investigated [6–8,12,18–21] to improve registration performance, especially efficiency. Among existing works, unsupervised learning is currently the most popular way

© The Author(s), under exclusive license to Springer Nature Switzerland AG 2022
S. Ali et al. (Eds.): CaPTion 2022, LNCS 13581, pp. 159–168, 2022.
https://doi.org/10.1007/978-3-031-17979-2_16

to train the registration network [2, 4, 5, 11] since it does not need ground-truth deformations, which is difficult to obtain in practice. Basically, the input is the intensity image pair, and the loss function is defined by two main terms: 1) the image similarity between the to-be-registered image pair and 2) the regularization to constrain the smoothness of the deformation field, which is also the final output of the network. Although these methods can partially improve the registration performance, it is still challenging when the images have large local deformations. In this case, the smoothness is even hard to preserve.

A typical clinical scenario is adaptive radiation therapy (ART) [22] for prostate and cervical cancer. To make the radiotherapy more precise, it is necessary to align the planning CT to treatment CT so that the dose planning can be well adaptive to the treatment status. However, the bladder and rectum easily have large local deformations for pelvic CT images, making the registration difficult in the actual application. To largely avoid the side effect and make the treatment more precise, it is essential to accurately align the images with sizeable local deformation while keeping the topology consistent for organs and clinical target volume.

In order to enhance registration accuracy for some specific organs, some works [10, 13, 14, 23] have used organ segmentations as the auxiliary information to train the registration model. In these works, organ segmentation and original intensity images are input to the network, and additional loss can be defined by the overlay of organ segmentations (e.g., Dice loss). The large deformation of specific organs can be more easily estimated by organ segmentation. But some limitations need to be further investigated. First, the smoothness of these organs and the topology consistency are difficult to preserve, especially for the organs with large deformations. Second, the model is trained by using a specific organ mask, when changing to other organs, the model needs to retrain from scratch, which is not flexible.

Some works [9, 17] have begun to pay attention to the anatomical structure of organs and use anatomical constraints to get more accurate registration results. However, no work has focused on the preservation of organ topology. Topology preservation is a common problem for registration tasks. When performing registration, the intensity images are often processed as a whole. The intensity difference or organ segmentation are often used to distinguish different organs. Basically, the morphology or structure of different anatomies is different. If some semantic information can be used to represent organ structure and incorporated into the registration model by adding the topology constraint, the challenging registration problem can be effectively improved.

The human anatomy can be expressed by some typical structures, as shown in Fig. 1. (1) Volume-structure. In pelvic images, the bladder and rectum can be regarded as volume structure, the semantic information can be defined by a surface in 3D space or a contour in 2D space. (2) Tubular-structure. The vessels and bronchus are more like tubular structures. The semantic information of this kind of structure can be defined by a skeleton or centerline (i.e., a line structure point set).

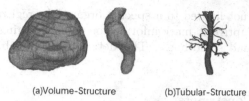

(a)Volume-Structure   (b)Tubular-Structure

Fig. 1. Some typical structures for different anatomies. (a) bladder and rectum; (b) vessels.

This paper proposes a novel semantic-aware registration framework to accurately align the organs with large local deformation for pelvic CT images and keep the topology smooth and consistent. It will improve the registration model by leveraging the semantic information to represent the organ morphology. The main contributions of this work are as follows:

(1) The registration network is designed by introducing semantic information to comprehensively improve registration performance. The extracted semantic information is simple but enough to represent complex organ morphology.
(2) We introduce additional regularization constraints based on semantic information, i.e., organ surface and contour during the model training. The sizeable local deformation can be effectively estimated, and meanwhile, the organ topology can also be well preserved.
(3) We present a flexible training framework by randomly using organ masks during training, not limited to a specific organ. In the inference stage, the input organ mask can be changed to any other organs without additional training, which can be extended to fulfill different requirements.

2   Method

The main task of deformable registration is to optimize the deformation field ϕ, which mapping coordinates from one image to another (e.g., fixed image I_f and moving I_m), so that the warped moving image $T(I_m, \phi)$ is similar to fixed image I_f. $T()$ is the spatial transformation based on ϕ. The energy function can be defined as:

$$E(\phi) = argminL_D(I_f, T(I_m, \phi)) + \lambda L_R(\phi) \tag{1}$$

where L_D denotes the dissimilarity metric between images after registration. $L_R(\phi)$ denotes the regularization term to maintain the smoothness of the deformation field ϕ.

There are two main differences compared with conventional learning-based registration. (1) We design a novel regularization loss, i.e., the **structural constraints**, based on the semantic information, e.g., the surface and contour of an organ, to preserve the organ topology. Even with sizeable local deformation, the implausible distortion can still be avoided. (2) The **adaptive training** framework is proposed. The input includes both the intensity and organ mask. The mask is not limited to a specific organ. The registration model can automatically be adaptive to mask information when changing the organ mask, ensuring accuracy and smoothness. The details will be elaborated on in the following sections.

2.1   Structural Constraints

For large volume organs, surface and contour can well present the organ morphology. Thus, the smoothness of the surface is crucial to preserve the organ topology. Conventional smoothness constraint is applied to the entire deformation field. Additional smoothness constraints should also be considered to register the organs with large deformation. Here, the coordinate of a point i is (x_i, y_i, z_i), the corresponding deformation vector field (DVF) is $(\Delta x^i, \Delta y^i, \Delta z^i)$.

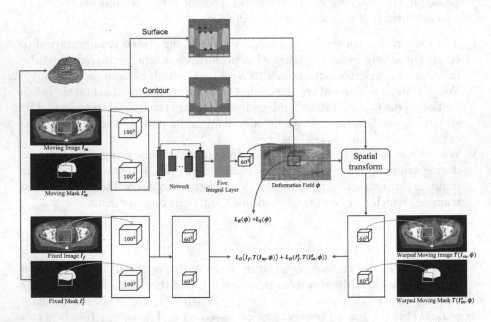

Fig. 2. The framework of proposed learning-based image registration. (Color figure online)

Surface Constraint. To keep the organ surface smooth after registration, the adjacency relation of the vertexes on the surface should keep the same. The DVF on the surface map should be minimized, and the deformation can consequently be consistent in the local regions. The implementation is shown in Fig. 2. The surface is constructed from segmentation mask by Marching Cubes algorithm [16]. The reconstructed surface is a mesh structure consisting of many triangular facets. To keep the surface smooth, for an arbitrary vertex (the red vertex in Fig. 2), the adjacent vertexes (the purple vertexes in Fig. 2) should keep the same after registration. Mathematically, it can be formulated by:

$$R_s = \frac{\sum_{j \in S} \sum_{k \in A_j} |v^j - v^k|}{\sum_{j \in S} \sum_{k \in A_j} 1} \tag{2}$$

where j denotes an arbitrary vertex on the mesh. k denotes the adjacent vertex of j. S denotes all vertexes on the surface mesh. A_j denotes the set of all adjacent vertexes of point j. v^j, v^k denotes the DVF of j and k respectively. $v^j = (\Delta x^j, \Delta y^j, \Delta z^j)$, $v^k = (\Delta x^k, \Delta y^k, \Delta z^k)$.

Contour Constraint. This constraint is designed when the images have large slice thickness. The main idea of contour constraint is to maintain the topological structure within slices and avoid unnecessary interpolation across slices. After the registration, the points within one slice should be consistent. Mathematically, it can be formulated by:

$$R_c = \frac{\sum_{l \in L} \sum_{i \in l} (v_z^i - \mu_l)^2}{\sum_{l \in L} \sum_{i \in l} 1} \tag{3}$$

where L is a super-set, and its subset l contains a series of vertexes i on the same layer of surface mesh. μ_l is the average of the z axis DVF values of all vertexes on l. $v_z^i = (\Delta z^i)$

As a summary, the structural constraints loss can be designed as:

$$L_S(\phi) = \alpha R_s + \beta R_c \tag{4}$$

where α and β represent the weight of R_s and R_c respectively. In our experiments, we empirically set $\alpha = 1$, $\beta = 0.5$.

2.2 Adaptive Registration

Figure 2 shows the training pipeline of our proposed semantic-aware registration network. The input patch size is 100^3 and the output patch size is 60^3. The purpose of this is to use the surrounding information to guide the registration of the central area to obtain more accurate registration results. The network architecture is based on U-Net, similar to VoxelMorph [3]. We also use integration layer to keep the diffeomorphic property during training.

Randomly Mask Input. We input pairs of images and masks into the network in the training stage. Here, the mask is not limited to specific organs. It is randomly selected in each training iteration (in this work, we use bladder and rectum masks). The random mask input can make the network perceive more mask information and avoid overfitting to a specific organ, which is more flexible. In the inference stage, based on the requirements, any organ mask can be used as input to enhance the registration performance, including smoothing and accuracy, without training a new model from scratch.

Above all, to train the proposed semantic-aware registration network, the entire loss function can be formulated by:

$$L = L_D(I_f, T(I_m, \phi)) + \lambda_1 L_D(I_f^s, T(I_m^s, \phi) + \lambda_2 L_R(\phi) + \lambda_3 L_S(\phi) \qquad (5)$$

$L_D(I_f, T(I_m, \phi))$ denotes MSE loss between fixed image and moving image, $L_D(I_f^s, T(I_m^s, \phi)$ denotes DICE loss between fixed mask and moving mask, $L_R(\phi)$ denotes the regularization term for the entire deformation field (often defined by the function of spatial gradients of ϕ), $L_S(\phi)$ denotes the topological constraint term proposed in this paper. λ is the weight of each term. After many experiments, we conclude that the model performs well when $\lambda_1 = 16$, $\lambda_2 = 6$, $\lambda_3 = 5$. Each network was trained for 1500 epochs, using ADAM optimizer with the learning rate of $5 * 10^{-6}$.

3   Experiments

Table 1. The test results comes from 25 patients with cervical cancer or prost ate cancer, a total of 92 pairs of CT images. The evaluation indicators include Dice similarity coefficient (DSC), symmetric average surface distance (SASD) and Hausdorff distance (HAUS). Bold indicates that the current item performs best in the test set. DF demotes diffeomorphic registration [4], MA denotes mask.

	Method	Bladder	Rectum
Avg. DSC (%)	ANTs-SyN	75.5 ± 13.9	80.2 ± 7.4
	U-Net	74.1 ± 13.6	73.7 ± 9.2
	U-Net+DF	74.2 ± 13.7	74.5 ± 8.7
	U-Net+DF+MA	92.3 ± 5.8	84.1 ± 4.6
	Proposed	$\mathbf{93.3 \pm 4.7}$	$\mathbf{86.6 \pm 3.7}$
SASD (mm)	ANTs-SyN	2.8 ± 3.2	0.4 ± 0.9
	U-Net	2.3 ± 3.4	0.5 ± 0.7
	U-Net+DF	2.4 ± 3.3	0.6 ± 1.0
	U-Net+DF+MA	0.6 ± 1.4	0.4 ± 0.6
	Proposed	$\mathbf{0.5 \pm 1}$	$\mathbf{0.3 \pm 0.5}$
HAUS (mm)	ANTs-SyN	9.8 ± 13.5	2.2 ± 4.1
	U-Net	9.9 ± 13.3	1.8 ± 2.4
	U-Net+DF	9.7 ± 13.2	2.2 ± 3.4
	U-Net+DF+MA	4.0 ± 7.3	2.3 ± 3.7
	Proposed	$\mathbf{3.8 \pm 6.1}$	$\mathbf{1.6 \pm 3.0}$

Our experimental data comes from 112 patients with prostate cancer or cervical cancer. Each patient has two to three pelvic CT scans, a total of 346 pairs of CT images. A 254 pair of images from 87 patients is used for training sets, and 92

Table 2. The test results comes from 25 patients with cervical cancer or prostate cancer, a total of 92 pairs of CT images. The evaluation indicators include the number of $|J_\phi| < 0$ and registration time (in second). Bold indicates that the current item performs best in the test set. DF demotes diffeomorphic registration [4], MA denotes mask.

	Method	Avg.	Std.
$|J_\phi| < 0$	ANTs-SyN	0	0
	U-Net	179271	92934
	U-Net+DF	1386	1373
	U-Net+DF+MA	1951	1949
	Proposed	**1114**	**1123**
Avg. registration time (s)	ANTs-SyN	180	5
	U-Net	**13**	**4**
	U-Net+DF	15	4
	U-Net+DF+MA	20	3
	Proposed	20	4

from 25 patients are used for test sets. The radiologist has manually delineated the bladder and rectum for each pelvic CT scan. All to-be-registered image pairs have already performed rigid registration (FLIRT [15]). The spacing and image size are $(1 * 1 * 3 \text{ mm}^3)$ and $(512 * 512 * 82)$. We use a single NVIDIA TITAN RTX graphics card when testing the registration time.

In the training stage, the inputs are the fixed and moving image pair with corresponding masks (totally four channels). The training patch size is 100 * 100 * 100, and the center point of the patch is randomly selected. The mask sent to the network each iteration is also randomly selected, and different masks correspond to different constraints.

We use four metrics to evaluate the registration performance: 1) Dice similarity coefficient (DSC), 2) symmetric average surface distance (SASD), 3) Hausdorff distance (HAUS), and 4) $|J_\phi| < 0$, the number of non-zero elements of the Jacobi matrix. The registration time is also evaluated to measure the efficiency of different methods. We compare our proposed method with the state-of-the-art registration method (i.e., ANTs-SYN [1]) and some typical deep learning-based registration frameworks [4]. We use the same network architecture for deep learning-based methods to make the experimental results comparable.

3.1   Registration Results

Table 1 lists the average DSC, SASD and HAUS of various methods. As can be seen from Table 1, the registration accuracy of our proposed method is optimal for different organs. In addition, Table 2 lists the number of $|J_\phi| < 0$ and registration time of different methods. As can be seen from Table 2, the number of $|J_\phi| < 0$ for ANTS [1] method is 0, which means that ANTS [1] can well preserve the topological consistency of the organ. However, the registration time of ANTS [1] is very long, and its registration results for large deformation organs are not accurate enough (see Table 1). Among the learn-based methods, our method is optimal under the evaluation of $|J_\phi| < 0$. In the average registration time for all methods, a simple U-net network is the fastest, which is due to the fact that it has only two channels of input. Although our proposed network has 4-channel input, but this does not increase the registration time significantly.

Figure 3 shows the visible registration results of bladder. It can be seen that the registration results of ANTs and U-net without mask guidance are not accurate. With the help of mask information, the large deformation of bladder can be estimated more accurately (U-net+mask). However, the topology of the organ still not well preserved, as we can see some distortions on the organ surface. After adding our proposed topological constraint term, the bladder can be registered accurately, and the smoothness of the organ surface is also preserved well. In summary, our proposed method can improve the registration accuracy and preserves the organ topology.

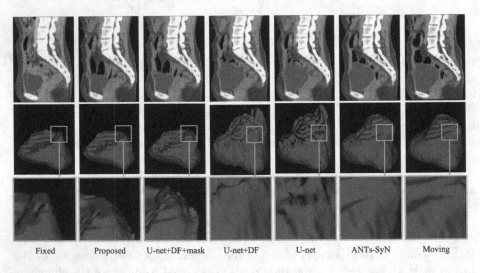

| Fixed | Proposed | U-net+DF+mask | U-net+DF | U-net | ANTs-SyN | Moving |

Fig. 3. The registration results are shown. The first line is the CT image, and the second line is the 3D reconstruction display of the registered mask. The third line shows the local magnification information of the 3D reconstruction.

4   Conclusion

In this paper, we propose a semantic-aware registration framework that uses the semantic information of organs to train the registration network, ensuring organ surface smoothness while improving registration accuracy. A flexible training strategy guarantees the robustness of the model, making the model flexible to multiple organ registration tasks. Compared with the traditional methods, our method can achieve significant improvements in terms of both registration accuracy and smoothness.

References

1. Avants, B.B., Epstein, C.L., Grossman, M., Gee, J.C.: Symmetric diffeomorphic image registration with cross-correlation: evaluating automated labeling of elderly and neurodegenerative brain. Med. Image Anal. **12**(1), 26–41 (2008)
2. Balakrishnan, G., Zhao, A., Sabuncu, M.R., Guttag, J., Dalca, A.V.: An unsupervised learning model for deformable medical image registration. In: Proceedings of the IEEE Conference on Computer Vision and Pattern Recognition, pp. 9252–9260 (2018)
3. Balakrishnan, G., Zhao, A., Sabuncu, M.R., Guttag, J., Dalca, A.V.: VoxelMorph: a learning framework for deformable medical image registration. IEEE Trans. Med. Imaging **38**(8), 1788–1800 (2019)
4. Dalca, A.V., Balakrishnan, G., Guttag, J., Sabuncu, M.R.: Unsupervised learning for fast probabilistic diffeomorphic registration. In: Frangi, A.F., Schnabel, J.A., Davatzikos, C., Alberola-López, C., Fichtinger, G. (eds.) MICCAI 2018. LNCS, vol. 11070, pp. 729–738. Springer, Cham (2018). https://doi.org/10.1007/978-3-030-00928-1_82
5. De Vos, B.D., Berendsen, F.F., Viergever, M.A., Sokooti, H., Staring, M., Išgum, I.: A deep learning framework for unsupervised affine and deformable image registration. Med. Image Anal. **52**, 128–143 (2019)
6. Eppenhof, K.A., Lafarge, M.W., Moeskops, P., Veta, M., Pluim, J.P.: Deformable image registration using convolutional neural networks. In: Medical Imaging 2018: Image Processing, vol. 10574, pp. 192–197. SPIE (2018)
7. Hansen, L., Heinrich, M.P.: GraphRegNet: deep graph regularisation networks on sparse keypoints for dense registration of 3D lung CTs. IEEE Trans. Med. Imaging **40**(9), 2246–2257 (2021)
8. Heinrich, M.P.: Closing the gap between deep and conventional image registration using probabilistic dense displacement networks. In: Shen, D., et al. (eds.) MICCAI 2019. LNCS, vol. 11769, pp. 50–58. Springer, Cham (2019). https://doi.org/10.1007/978-3-030-32226-7_6
9. Hering, A., Häger, S., Moltz, J., Lessmann, N., Heldmann, S., van Ginneken, B.: CNN-based lung CT registration with multiple anatomical constraints. Med. Image Anal. **72**, 102139 (2021)
10. Hoffmann, M., Billot, B., Iglesias, J.E., Fischl, B., Dalca, A.V.: Learning multimodal image registration without real data (2020)
11. Hoopes, A., Hoffmann, M., Fischl, B., Guttag, J., Dalca, A.V.: HyperMorph: amortized hyperparameter learning for image registration. In: Feragen, A., Sommer, S., Schnabel, J., Nielsen, M. (eds.) IPMI 2021. LNCS, vol. 12729, pp. 3–17. Springer, Cham (2021). https://doi.org/10.1007/978-3-030-78191-0_1
12. Hu, X., Kang, M., Huang, W., Scott, M.R., Wiest, R., Reyes, M.: Dual-stream pyramid registration network. In: Shen, D., et al. (eds.) MICCAI 2019. LNCS, vol. 11765, pp. 382–390. Springer, Cham (2019). https://doi.org/10.1007/978-3-030-32245-8_43
13. Hu, Y., et al.: Label-driven weakly-supervised learning for multimodal deformable image registration. In: 2018 IEEE 15th International Symposium on Biomedical Imaging (ISBI 2018), pp. 1070–1074. IEEE (2018)
14. Hu, Y., et al.: Weakly-supervised convolutional neural networks for multimodal image registration. Med. Image Anal. **49**, 1–13 (2018)
15. Jenkinson, M., Smith, S.: A global optimisation method for robust affine registration of brain images. Med. Image Anal. **5**(2), 143–156 (2001)

16. Lorensen, W.E., Cline, H.E.: Marching cubes: a high resolution 3D surface construction algorithm. ACM SIGGRAPH Comput. Graph. **21**(4), 163–169 (1987)
17. Mansilla, L., Milone, D.H., Ferrante, E.: Learning deformable registration of medical images with anatomical constraints. Neural Netw. **124**, 269–279 (2020)
18. Mok, T.C., Chung, A.: Fast symmetric diffeomorphic image registration with convolutional neural networks. In: Proceedings of the IEEE/CVF Conference on Computer Vision and Pattern Recognition, pp. 4644–4653 (2020)
19. Mok, T.C.W., Chung, A.C.S.: Large deformation diffeomorphic image registration with Laplacian pyramid networks. In: Martel, A.L., et al. (eds.) MICCAI 2020. LNCS, vol. 12263, pp. 211–221. Springer, Cham (2020). https://doi.org/10.1007/978-3-030-59716-0_21
20. Mok, T.C.W., Chung, A.C.S.: Conditional deformable image registration with convolutional neural network. In: de Bruijne, M., et al. (eds.) MICCAI 2021. LNCS, vol. 12904, pp. 35–45. Springer, Cham (2021). https://doi.org/10.1007/978-3-030-87202-1_4
21. Rohé, M.-M., Datar, M., Heimann, T., Sermesant, M., Pennec, X.: SVF-Net: learning deformable image registration using shape matching. In: Descoteaux, M., Maier-Hein, L., Franz, A., Jannin, P., Collins, D.L., Duchesne, S. (eds.) MICCAI 2017. LNCS, vol. 10433, pp. 266–274. Springer, Cham (2017). https://doi.org/10.1007/978-3-319-66182-7_31
22. Yan, D., Vicini, F., Wong, J., Martinez, A.: Adaptive radiation therapy. Phys. Med. Biol. **42**(1), 123 (1997)
23. Zhu, W., et al.: NeurReg: neural registration and its application to image segmentation. In: Proceedings of the IEEE/CVF Winter Conference on Applications of Computer Vision, pp. 3617–3626 (2020)

Author Index

Printed in the United States
by Baker & Taylor Publisher Services

Printed in the United States
by Baker & Taylor Publisher Services